T0257986

Cryopreservation: Modern Insights

Cryopreservation: Modern Insights

Edited by **Marianne Wilde**

New York

Published by Callisto Reference,
106 Park Avenue, Suite 200,
New York, NY 10016, USA
www.callistoreference.com

Cryopreservation: Modern Insights
Edited by Marianne Wilde

International Standard Book Number: 978-1-63239-135-3 (Hardback)

Printed in the United States of America.

Contents

Preface

After a long gap of almost a decade, here is a book on cryopreservation. Lately, there have been certain severe tectonic shifts in cryobiology though not visible on the surface but will have significant impact on both the advancement of novel cryopreservation techniques and the future of cryobiology. This comprehensive book discusses the existing applications and emerging practical protocols for the purpose of cryopreservation along with the description of the novel cryobiological ideas. The topics have been divided into the following sections namely, Cryopreservation of Aquatic Species, Cryopreservation in Plants, and Equipment and Assays.

This book is the end result of constructive efforts and intensive research done by experts in this field. The aim of this book is to enlighten the readers with recent information in this area of research. The information provided in this profound book would serve as a valuable reference to students and researchers in this field.

At the end, I would like to thank all the authors for devoting their precious time and providing their valuable contribution to this book. I would also like to express my gratitude to my fellow colleagues who encouraged me throughout the process.

Editor

Part 1

Cryopreservation of Aquatic Species

Sperm Cryopreservation of Some Freshwater Fish Species in Malaysia

Poh Chiang Chew and Abd. Rashid Zulkafli
Freshwater Fisheries Research Division, FRI Glami Lemi, Jelebu,
Negeri Sembilan,
Malaysia

1. Introduction

It has been estimated that spermatozoa can last from 200-32,000 years (Stoss & Donaldson, 1983; Suquet et al., 2000). According to Kopeika et al. (2007), several methods of fish sperm storage has been practised including stored in medium with saturated gases, preservation at temperature above zero, in a frozen state as well as by drying. However, the low temperature approaches have been successful in fish sperm cryopreservation. Thus, cryopreservation technology offers the best means for long term storage of fish semen.

To date, successful cryopreservation of fish semen were reported in more than 200 freshwater species and 40 marine species worldwide (Gwo, 2000). Even though in general many successes have been achieved in fish semen cryopreservation, the technique remains as a method that is difficult to be standardized and use in all types of fishes. This is due to the fact that cryopreservation of sperms from different fish species required different conditions, where the protocol needs to be established individually. Even the "general protocol" of cryopreservation of fish sperm summarized by Kopeika et al. (2007) encompassed many variations when different species of fish are involved, particularly in the use of medium ingredients for the cryopreservation.

In view of need to develop individual protocol for successful cryopreservation of fish semen and considering Malaysia has a rich fish fauna with many of them unique to this tropical region, cryopreservation of fish gametes will require detailed study to create new protocol for each fish species intended for semen cryopreservation. To date in Malaysia semen cryopreservation has only been reported for several freshwater fish species, namely *Probarbus jullieni*, *Tor tambroides*, *T. deuronensis*, *Hemibagrus nemurus*, *Pangasius nasutus*, *Hypsibarbus wetmorei*, *Barbonymus gonionotus* and *Clarias gariepinus*. It has been demonstrated that semen cryoperservation plays an important role for the genetic conservation of these fish species.

Cryopreservation technology for fish semen is still not well explored in Malaysia and can be considered as new if compared to the domesticated terrestrial livestocks. Henceforth, this has opened up a new field to be explored with potential applications in aquaculture and in the conservation of the national fisheries genetic resources. Cryopreserved semen could facilitate artificial fertilization especially when mature male fishes are not available or unable to provide viable semen during certain periods of the breeding season. Semen cryopreservation may also be useful for fertilization to produce hybrids of various fish species. It also helps in reducing the cost and labor of maintaining broodstocks under *in situ* condition. In line with the mission of Department of Fisheries (DOF) Malaysia to develop and manage the national fisheries sector in a sustainable manner, the gene bank of freshwater fishes in the form of semen cryobank of Fisheries Research Institute at Glami Lemi was established in 2008. The establishment of the semen cryobank research has achieved the aim of the DOF towards establishing a national semen cryobank (gene bank) in Malaysia for conserving the genetic materials of the threatened or endangered indigenous freshwater fish species and also for those indigenous species which has potential for aquaculture.

The main focus of this chapter will be on the methodology developed for the semen cryopreservation in Malaysia of some fish species mentioned above and the various important steps and several key factors that contributed to the successes in gamete cryopreservation. In addition, the chapter will also present the current status and the challenges of fish semen cryopreservation in Malaysia, especially on the conservation of genetic resources and potential applications of gamete cryopreservation in aquaculture. Challenges in establishment and maintenance of the fish sperm cryobank are also discussed.

2. Semen cryopresevation of freshwater fish species of Malaysia

Malaysia has close to a hundred river systems, two natural lakes (Lake Chini and Lake Bera) and a vast freshwater water bodies and peat swamps (Baluyut, 1983). On fish biodiversity, Malaysia has approximately 616 reported freshwater fish species (Froese & Pauly, 2003). Most of the inland fisheries resources are dominated by the cyprinids and silurids (Khoo et al., 1987). Some endemic species are found in rivers, lakes and peat swamps. In the past decades intensive development activities such as deforestation and land clearing for oil palm plantation or other agricultural uses, indiscriminate logging within and around the catchment areas and construction of dams for hydroelectricity, has led to many riverine fish species suffer high risk of extinction (Khoo et al., 1987; Jackson & Marmulla, 2001). These development activities have resulted in habitat destruction, deterioration of water quality, water pollution and sedimentation, especially during rainy season when runoff is increased. All these certainly have direct great impacts on some fish species and the impacts are irreversible (Ho, 1995). Apart from the environmental degradation resulting from development, other causes of loss in inland fisheries biodiversity are attributed to over-exploitation due to strong market demand, the use of illegal and destructive fishing gears such as poisoning and electro-shocker and the introduction of invasive exotic species (Dudgeon, 2002; Allan et al., 2005). The indigenous species such as Isok barb (*Probarbus jullieni*), Malaysian Mahseer (*Tor* spp.), Mad barb (*Leptobarbus hoeveni*), Hampala barb (*Hampala macrolepidota*), knife fish (*Chitala lopis*), climbing perch (*Anabas testudineus*), snakeheads (*Channa* spp.), Asian arowana (*Scleropages formosus*), the pangasiid catfishes (*Pangasius nasutus, Pangasigodon waandersii*), giant river catfish (*Wallago leerii*), large headed walking catfish (*Clarias macrocephalus*) and giant gouramy (*Osphronemus goramy*) have dwindled in great numbers continuously due to unsustainable fishing activities. At present, most Malaysian fish species could only be conserved probably in the inaccessible or remote areas of the country.

It is no doubt that continuing habitat destruction, overfishing and competition for food from the invading fish species are leading to loss of inland fisheries biodiversity even before much of them could be documented. The lack of data on the indigenous species will subsequently impede efforts to better utilize and manage the nation's inland fisheries resources in a sustainable manner, and can eventually resulting in threatened, endangered or extinction of fish species in Malaysia. With respect to National Biodiversity Policy, it is therefore vital to protect and safeguard the indigenous fisheries resources while the species still exist in the wild. Realizing the risk of depleting fish stocks from natural waters, studies in domestication, management and husbandry of broodstocks, artificial breeding, grow out, nutrition and effort in stock enhancement via restocking of selected indigenous freshwater fish species have been carried out by the Department of Fisheries, Malaysia ever since 1980's. The species which have been studied and documented were the indigenous catfishes (*Clarias macrocephalus* and *C. batrachus*), Tropical bagrid catfish (*Hemibagrus nemurus*), Hampala barb (*Hampala macrolepidota*), Isok barb (*Probarbus jullieni*), Malaysian Mahseer (*Tor tambroides*), local pangasiid catfish (*Pangasius nasutus*) (Pathmasothy & Omar, 1982; Pathmasothy, 1985; Saidin, 1986; Thalathiah et al., 1988; Mohamad-Zaini, 1992; Thalathiah et al., 1992; Ahmad-Ashhar, 1992; Ahmad-Ashhar & Haron, 1994; Ahmad-Ashhar, 1996; Suhairi, 1996; Suhairi et al., 1996; Ahmad-Ashhar, 1998).

In 2007, cryopreservation of semen was implemented as one of the *ex situ* conservation approach, with the exotic species such as the Javanese barb (*Barbonymus gonionotus*) and African catfish (*Clarias gariepinus*) used as the model species to examine the various suitable formulations diluents and skill development in semen cryopreservation. These two species were chosen because they are domesticated species, which are available abundantly in Malaysia and able to breed easily in captivity. In Malaysia, very little work was done on semen cryopreservation in fish species and study on fish semen cryopreservation is still at its infancy. So far only a few indigenous fish species, namely the *P. jullieni*, *Tor* spp, *H. nemurus*, *P. nasutus* and *H. wetmorei* were studied. Of which only studies on the *P. jullieni*, *Tor* spp., and *H. nemurus* were reported (Chew et al., 2010a; Chew et al., 2010b; Muchlisin et al., 2004), while studies on other species remain unpublished. The biology, ecology and reproduction biology of these indigenous fish species are described in the following:

2.1 *Probarbus jullieni*

P. jullieni (English name: Isok barb or Jullien's Golden Carp or seven-striped barb) of family Cyprinidae (Figure 1a) is commonly known as Temoleh or Temelian among Malaysians. This species is listed in the Convention on International Trade in Endangered Species of Wild Fauna and Flora (CITES), Appendix 1 and the IUCN Red List as the endangered species (Hogan et al., 2009; IUCN, 2009). Therefore, conservation of this species is in urgent needed. Isok barb is a migratory species inhabiting river channels with water depth more than 10 m and is endemic to the Pahang River and Perak River in Peninsular Malaysia and the Mekong River basins of Indochina (Roberts, 1992). However, the drastic depletion of the numbers of Isok barb in Perak River was reported ever since the construction of Chenderoh Dam in 1930s. The dam created a physical barrier at the Perak River that permanently blocked the migration path of the Isok barb and increased water level further destroyed the spawning ground of this species (Khoo et al., 1987). Although the Isok barb can grow to reach the weight exceeded 70 kg, like the size of a human being (Baird, 2006), the landing of large size Isok barb (>10kg) has rarely occurred in the past 10 years.

Isok barb is a seasonal-bred species. In Malaysia, induced breeding of this species was carried out successfully in FRI Glami Lemi (formerly known as Freshwater Fisheries Research Centre) in 1990s (Ahmad-Ashhar, 1992; Ahmad-Ashhar & Haron, 1994). In its natural environment, this species spawn in fast flowing deep waters with sandy bottom. Chew et al. (2010a) reported that the spawning behavior of *P. jullieni* in captivity very much associated with the monsoons. Nevertheless, the breeding season of the cultured Isok barb was reported to be 1-2 months earlier (October to January) than the wild populations (December to February).

2.2 *Tor* spp. (*T. tambroides* and *T. deuronensis*)

The Malaysian Mahseer, *Tor* spp. (family: Cyprinidae) or locally known as Kelah, Empurau, Semah or Pelian, is highly sought for its value as a food fish, game fish and ornamental fish (Inger & Chin, 1962; Mohsin & Ambak, 1991; Kottelat et al., 1993). Two valid *Tor* species were identified and described in Malaysia, i.e. *T. tambroides* and *T. douronensis* (Roberts, 1989; Kottelat et al., 1993; Rainboth, 1996; Zhou & Chu, 1996) (Figure 1b and 1c). Of the two *Tor* spp., the *T. tambroides* is more favourable as food fish and can fetch up to about USD 100 – USD 220 per kg and thus is the most expensive freshwater food fish in Malaysia. Malaysian Mahseer occurs in most undisturbed and clear flowing upstream rivers, reservoir systems and lotic habitats throughout the East and Southeast Asia. The major geographical locations of Malaysian Mahseer are Pahang, Perak, Terengganu, Kelantan, Sabah and Sarawak (Ng, 2004; Ambak et al., 2007). Similar to other indigenous fishes in Malaysia, the wild populations of Malaysian Mahseer are declining over the years as the consequences of over-exploitation, natural habitat degradation and water pollution. Therefore, Malaysian Mahseer is now classified nationally as ecologically threatened (Ingram et al., 2005).

The taxonomic status of species within the genus *Tor* has been highly contentious due to plasticity of many external morphological features resulted from considerable geographical and ecological variability (Tsigenopoulos & Berrebi, 2000; Nguyen et al., 2007). Therefore, species identification based on morphological comparison sometimes could be confusing to most people who are not trained in fish taxonomy. Ability to differentiate between the Malaysian Mahseer and the Copper Mahseer (*Neolissochilus* spp.) is another difficult task hampered most of the fish farmers or aqua-culturists. The Copper Mahseer is not a desirable species and thus its value in term of market price is very much lower compared to the Malaysian Mahseer. At present, the seed supply of Malaysian Mahseer is still depending solely on the captive wild stock. As such, those who want to culture this species will have to bear with the risk of getting seed stock that comprises the mixture of these two species. Therefore, a more effective method for species identification by using DNA markers such as the mitochondria DNA (mtDNA) sequences and microsatellte markers are seen as potential in solving this problem (Nguyen et al., 2006a; Nguyen et al., 2006b; Nguyen et al., 2008; Nguyen, 2008).

The first successful hormonal induction spawning of *T. tambroides* under captive pond-reared environment was reported by Ingram et al. (2005). In captive condition, *T. tambroides* seems to spawn all year round. Brood fish of both sexes may strip up to a few times in a year. However, it is reported that geographical reproductive diversity, diet and environment conditions such as changes in temperature, water level, pH, flow velocity, turbidity, rain falls, collectively trigger the Mahseer to spawn (Dobriyal et al., 2000). Malaysian Mahseer is also a large riverine species and that can grow up to 30 – 50kg. In the culture condition, the growth rate of Malaysian Mahseer is relatively slow compared to other aquaculture species. This species usually took up to three years to reach the marketable size of 1.5 – 2.0 kg.

2.3 *Hemibagrus nemurus*

The tropical bagrid catfish, *H. nemurus* (previously known as *Mystus nemurus*) of family Bagridae and with the vernacular name Baung (Figure 1d), is a widely distributed food fish found in most of the inland water bodies in Malaysia (Khan et al., 1995). The occurrence of this species in brackish water was also reported (Inger, 1955). The tropical bagrid catfish is a bottom feeder and fed on a wide range of food from small teleosts, crustaceans, benthic invertebrates to the detritus (Khan, 1987). This species is potential for aquaculture as it receives good acceptance from the local market due to its tender and non-bony meat. This carnivorous species is also a popular species in sport fishing. The artificial spawning of tropical bagrid catfish via hormonal induction using the heteroplastic pituitary extract in combination with human chorionic gonadotropin (HCG) was reported by Thalathiah et al. (1988, 1992). According to Khan et al. (1995), a bimodal male and female gonad somatic index (GSI) pattern throughout a year was shown in this species, suggesting this species could spawn twice in a year.

2.4 *Pangasius nasutus*

The local pangasiid catfish, *P. nasutus* of family Pangasiidae and with local name Patin Buah, (Figure 1e) is a native river catfish endemic to Pahang River in Malaysia. As commented by Hung et al. (2004), catfishes from the family Pangasiidae are of great economical importance in Southeast Asia. *P. nasutus* is one of the favorite food fish due to its white and tender flesh. This omnivorous species commands a high market value (USD 20 per kg) in local market. In the wild, *P. nasutus* fed mainly on the bivalves, mollusks, gastropods and other benthic organisms especialy on sandy bottom rivers (Roberts & Vidthayanon, 1991). Because of its economic value and declining population in the natural waters, this has created popularity and awareness to conserve and culture this species. The first successful induced spawning of *P. nasutus* was reported in 2005. This species is also a batch spawner and exhibits a single-modal GSI pattern throughout the whole year. During the spawning season, multiple releases of eggs from a single female broodfish were observed during the single period (Mohd-Zafri, 2006). The breeding season of this species is associated with rainfalls or monsoons between May – July (Maack & George, 1999).

2.5 *Hypsibarbus wetmorei*

Morphologically *Hypsibarbus wetmorei* is quite similar to the Javanese Barb, *Barbonymus gonionotus* except the appearance of turmeric-like yellow colour on its body (Figure 1f) where the vernacular name Kerai Kunyit is given. This species is of the family Cyprinidae. Kerai Kunyit is highly regarded as a food fish and thus it is a potential aquaculture species. In Malaysia, this species is found to be endemic in Pahang River. Until today, the biology, ecology and repoductive behaviour of this species is not well studied and limited documentation on the species is available. By far, only study on species identification of *Hypsibarbus* spp. (included *H. wetmorei*) using PCR-RFLP method was reported by Jantrarotai et al. (2007).

Fig. 1. Photos of the Malaysian fishes involved in the cryopreservation studies, (a) Isok barb (*P. jullieni*), (b) Malaysian Mahseer (*T. tambroides*), (c) Malaysian Mahseer (*T. deauronensis*), (d) Tropical bagrid catfish (*H. nemurus*), (e) Local pangasiid catfish (*P. nasutus*), and (f) Kerai Kunyit (*H. wetmorei*).

3. Methodology employed for semen cryopreservation of freshwater fishes in Malaysia

As in cryopreservation of other fish species, the cryopreservation of the semen of freshwater fishes from Malaysia to be discussed later falls under the following general procedures (i) collection of semen, (ii) dilution of semen sample, (iii) semen sample packaging, (iv) equilibration, (v) freezing, (vi) cryo-storage, and (vii) thawing. Successful cryopreservation of fish sperm very much depends on a range of factors in each step of the cryopreservation procedures as highlighted by Kopeika et al. (2007).

3.1 Collection of fish semen

Mature and healthy males broodfish were selected and anaesthesized using MS222 or clove oil according to the dosage recommended by the manufacturer. Semen sample was expelled from the male fish by gentle abdominal pressure and collected into a clean and dry tube. Extra precaution should be taken while collecting semen sample. Contamination of sample with blood, water, urine or the feces should be avoided as these contaminants significantly educed the semen quality and caused poor post-thaw sperm motility. The semen samples were then transferred back to laboratory for quantification of the fresh sperm quality and proceed with further dilution before freezing procedure. The sperm motility rates and sperm concentration of the freshly collected semen sample was evaluated prior to cryopreservation. In order to obtain good post-thaw motility, only semen samples showing sperm motility $\geq 70\%$ were used for cryopreservation.

3.2 Dilution of semen sample

The semen samples with good quality were subsequently diluted at an appropriate sperm to diluent ratio, with suitable extender solution and cryo-protectant. Sperm to diluent ratios ranged between 1:3 to 1:9 were reported to produce best results in fish sperm cryopreservation studies (Scott & Baynes, 1980; Lahnsteiner et al., 1996). Gwo (2000) reported the fish sperm could loss its viability in high dilution ratio especially in marine species. The type of extender solution, cryoprotectant and dilution ratio that were optimum for each Malaysian fish species studied were summarized in Table 1. The type of diluents and dilution ratios workable to preserve sperm motility appear to vary among different fish species. Thus each of these parameters needs to be optimized accordingly via a series of experimentations.

The extender solution helps to maintain sperm viability prior to and during the freezing process. Extender solution is a balanced salt buffer of specific pH and osmotic strength. Apart from salts, sometimes extender solution is prepared with addition of organic compounds such as glucose. The nature of the effect of extenders is based on the control of pH and salt concentration as well as the supply of energy, and can extend the functional life and fertilizing capability of the sperm (Tiersch, 2000). Cryoprotectants function to protect cells from cryo-damage or cryo-injury during freezing and thawing process. The permeating cryoprotectants, namely dimethyl sulfoxide, methanol, ethylene glycol and propylene glycol are among the most frequently used cryoprotective agents among the aquatic organisms (Lahnsteiner et al., 1997; Tiersch, 2000; Tiersch, 2006). However, the permeating cryoprotectants are often toxic to cells, and thus the choice of the types of cryoprotectant and their optimal concentration should be at a balance between protection and toxicity. On the other hand, the non-permeating cryoprotectants such as sucrose, glucose and polymers (e.g. alginate) were often used in combination with the extender solution in the diluents. Sometimes, a combination of different cryoprotectants in certain ratios could help improving the post-thaw motility. The studies of African catfish in our laboratory has shown that a combination of methanol and N, N-dimethylacetamide at ratio 70%:30% produced significant higher post-thaw motility compared to the use of a single cryoprotectant in sample dilution. Tiersch (2000) has also reported that the application of cryoprotectant at concentration between 5% to 20% ususally provides good protection in most fish species. The protective effect and optimal concentration of cryoprotectants could be species specific (Gwo et al., 1991; Suquet et al., 2000; Rideout et al., 2003). Therefore, the optimal concentration needs to be determined individually in each species studied through experimentations.

For Isok barb (*P. jullieni*), Malaysian Mahseer (*Tor* spp.), local pangasiid catfish (*P. nasutus*) and Kerai Kunyit (*H. wetmorei*), a total of 14 extender formulations, five types of cryoprotectants (dimethyl sulfoxide, ethylene glycol, glycerol, methanol and N,N - dimethylacetamide) with concentrations between 5-20% (v/v), semen to diluent ratios (1:1 to 1: 14) were examined as described by Chew et al. (2010a). This may be compared with Muchlisin et al. (2004) who used three extender solutions (the Ringer, physiological saline and saline solution); four types of cryoprotectants (DMSO, ethanol, glycerol and methanol) at three concentrations (5%, 10% and 15%) and three sperm to diluent ratios (1:20, 1:30 and 1:40) in tropical bagrid catfish sperm cryopresevation.

Species	Chemical composition of extender solution	Type of cryoprotectant & concentration (v/v)	Sperm to diluent ratio	Reference
B. gonionotus	Modified from Kurokura et al. 1984 - 62 mM NaCl, 134 mM KCl, 1.5 mM CaCl$_2$, 0.4 mM MgCl$_2$, 2.4 mM NaHCO$_3$ (Horvath et al., 2003)	Methanol 10%	1:4 – 1:6	Unpublished data
C. gariepinus	Ringer solution – 128 mM NaCl, 2.7 mM KCl, 1.4 mM CaCl$_2$, 2.4 mM NaHCO$_3$ (Kurokura et al., 1984)	Methanol + DMA (70:30) 8%	1:6	Unpublished data
P. jullieni	Calcium free Hank's Balance Salt Solution (CF-HBSS) – 152 mM NaCl, 5.9 mM KCl, 0.9 mM MgSO$_4$, 0.36 mM Na$_2$HPO$_4$, 0.5 mM KH$_2$PO$_4$, 4.6 mM NaHCO$_3$, 6.16 mM Fructose (Tiersch et al., 1994)	Methanol 9 - 10%	1:3 – 1:5	Chew et al., 2010a
Tor spp.	202 mM D(+)-glucose monohydrate, 51.5 mM sodium chloride and 6 mM sodium bicarbonate, with pH 7.1 to 7.8 and osmolality 309 $\pm$ 30 mOsmol/kg	DMSO 10%	1:7	Chew et al., 2010b
P. nasutus	CF-HBSS (Tiersch et al., 1994)	Methanol 9 - 10 %	1:7	Unpublished data
H. wetmorei	Modified Fish Ringer – 111 mM NaCl, 40.2 mM KCl, 2.1 mM CaCl$_2$, 2.4 mM NaHCO$_3$ (Wolf, 1963)	Methanol 9 - 10 %	1:4 – 1:7	Unpublished data
H. nemurus	Modified Fish Ringer – 128 mM NaCl, 2.7 mM KCl, 1.4 mM CaCl$_2$, 2.4 mM NaHCO$_3$, 25.3 mM glucose	Methanol 10%	1:20	Muchlisin et al., 2004

Table 1. A list of extender solution and its composition, type of cryoprotectant and its optimal concentration and sperm to diluent ratio for successful semen cryopreservation of various species of freshwater fishes in Malaysia.

3.3 Packaging and equilibration of diluted semen sample

In this procedure which is performed after the addition of extender solution and cryoprotectant, the diluted semen sample is packed into polyethylene (PE) straws (Chew et al., 2010a; Chew et al., 2010b) or cryo-vials (Muchlisin et al. ,2004). Extended semen sample is subjected to equilibration at temperature <10°C prior to freezing procedure. The duration taken for equilibration is the time required for the cryoprotectant to permeate the cells. Equilibration duration between 15 min to 3 h has been practised for Malaysian fish species sperm cryopreservation and was found to be successful in maintaining a good post-thaw sperm motility (Chew et al., 2010b). In most circumstances, the equilibration duration is set at 15 to 30 min, but it can be varied depending on the type and concentration of cryoprotectant used (Tiersch, 2000).

3.4 Sperm cryopreservation

Cryopreservation involves the removal of excess water from the inside of the cell to the exterior where it can form ice (Tiersch, 2000). A two-step rapid freezing method was used for semen cryopreservation in Isok barb, Malaysian Mahseer, local pangasiid catfish and Kerai Kunyit. After the equilibration procedure, semen samples of these species were subjected to liquid nitrogen vapour exposure for 5-10 min in an insulated styrofoam cooler box filled with liquid nitrogen, with the samples placed between 3 to 4 cm above the liquid nitrogen, after which the samples were directly plunged into liquid nitrogen. For tropical bagrid catfish (Muchlisin et al., 2004), the semen samples were equilibrated on crushed ice (0°C) for 5 min. After the equilibration procedure, the samples were then placed in an ice box containing dry ice (-120°C) for 5 min and subsequently plunged into liquid nitrogen.

Besides the styrofoam cooler and ice box, the vapour shipper is another choice of method could be used for the freezing of fish semen samples. To cryopreserve semen samples by using dry shipper, the extended and packed semen sample was transferred into the fully charged vapour shipper and subsequently transferred and stored in liquid nitogen. The use of the vapour shipper method in freezing as reported in *Tor* spp. (Chew et al., 2010b) was convenient to be used in the field. Such method is simple to use, the cooling rate is more consistent and controllable and it consumes less liquid nitrogen and requires less space compared to the styrofoam cooler box or ice box method.

In our opinion, the use of the sophisticated bench top type of controlled rate programmable freezer is not practical in cryopreservation of fish semen in some laboratories because of difficulty to move this equipment from one location to another. Unfortunately portable type of controlled rate programmable freezer is not user friendly, time consuming and limited number of samples could be processed at a single run. Compared with programmable freezer, the two-step freezing method employing insulated styrofoam cooler box or ice box is simpler, rapid and more cost effective. Nevertheless, the main shortcomings of this simple freezing method is the inconsistency in cooling rates and non-reproducible cryopreservation experiments if performed by different operators.

3.5 Thawing

Thawing is a process to recover the sperm cells from the dormant stage in ultra low temperature. Frozen semen is usually thawed at 40°C, with different optimal durations applied according to the type of packaging and the storage volume as shown in Table 2.

Type of packaging	Volume of semen	Duration of thawing
	0.25 mL	4s – 6s
PE straw	0.5 mL	6s - 8s
	1.2 mL	12s – 15s
Cryo-vial	5 mL	5 min

Table 2. The optimal durations used to thaw the frozen semen samples of freshwater fish species in Malaysia in different types of packaging and storage volume.

3.6 Egg fertilization

A cryopreservation protocol developed for a species is considered success if the semen that cryopreserved according to the certain protocol could successfully fertilize eggs and produced offspring. Fresh semen is usually included in the control treatment. The optimal sperm to egg ratio used should be determined prior to fertilization. The sperm: egg ratio at approximately 250,000:1 is usually sufficient and worked well in most species in our laboratory. However, other sperm to egg ratios from 1000:1 to 500000: 1 were tested by Butts et al. (2009) and 100 000: 1 was found to yield the best fertilization performance in Atlantic cod. The dry fertilization method is favorable in the fertilization test for freshwater fish and thus was applied for all Malaysian fish species studied in our laboratory. In the dry fertilization method, both eggs and sperms were mixed well before hatchery water was later added into the sperm/egg mixture to water-harden the fertilized eggs. After rinsing with hatchery water, fertilized eggs were then incubated in aquarium (with or without using a hatching trough depended on species). Water temperature was kept at room temperature between 23 - 28°C throughout the period of incubation. The duration of embryo development varied between species. Therefore, the duration required for the fertilized egg to hatch is also varied among different species.

4. Discussion on semen cryopreservation of Malaysian freshwater fish species

The successful semen cryopresevation of several Malaysian fish species e.g. *P. jullieni, T. tambroides, T. deuronensis, H. bleekeri, P. nasutus* and *H. wetmorei* using various methods of cryopresevation discussed above may be evaluated via analyses on the sperm motility, fertilization, hatching ability, etc.

4.1 Quality of semen

Sperm motility for the freshly collected semen from the healthy broodfish is usually ranged between 90 - 100% motility provided that the sperm sample was not activated prior to the actual cryopreservation processes or contaminated by water or urine while sampling was carried out. However from our observation, sperm motility of most freshwater fishes dropped to <10% after 24 h if the sample was not extended using suitable extender solution, even though the sample was kept cool in a refrigerator (0- 5°C).

Sperm concentration is one of the important characteristic that determines the sperm quality of an individual male (Billard, 1986; Suquet et al., 1992; Billard et al., 1995). The sperm concentrations of all Malaysian fish species studied were between 2.2×10^8 to 6.2×10^{10} sperm cells per mL and this is in agreement with the studies by Leung & Jamieson (1991). On the other hand, short lifespan after activation is the typical characteristic of spermatozoa of freshwater fish species. The duration of sperm motility of most freshwater fishes is normally <1 min after the sperms are activated (Billard & Cosson, 1992; Lahnsteiner & Patzner, 2008). Duration of sperm motility of *Tor* spp. was about 40-50 s and it slowed down drastically after 20 s of progressive movements (Chew et al., 2010b). For Isok barb, it was about 20 s and slowed down after 10 s of progressive movements. Table 3 shows the range of sperm concentrations and motility duration in each Malaysian species studied at FRI Glami Lemi.

Species	Sperm count (Number of cells per mL)	Duration of motility (*)
Javanese barb (B. gonionotus)	$2.52 \times 10^9 - 1.03 \times 10^{10}$	<15 sec (10 s)
African catfish (C. gariepinus)	$5.44 \times 10^9 - 1.14 \times 10^{10}$	15 – 30 s, (13 s)
Isok barb (P. jullieni)	$4.00 \times 10^8 - 6.24 \times 10^{10}$	20 s, (10 s)
Malaysian Mahseer (Tor spp.)	$2.20 \times 10^8 - 5.98 \times 10^9$	40 -50 s, (20 s)
Local pangasiid catfish (P. nasutus)	$6.60 \times 10^8 - 1.36 \times 10^{10}$	25 – 50 s, (15 s)
Kerai Kunyit (H. wetmorei)	$7.30 \times 10^9 - 1.33 \times 10^{10}$	20 -70 s, (20 s)

*Duration of progressive movement before slowed down and finally stopped

Table 3. Sperm concentration and motility duration of each freshwater fish species studied at FRI Glami Lemi.

Osmolality is another critical variable in sperm quality (Honeyfield & Krise, 2000). As seen in many studies, seminal plasma osmolality among males fish is highly variable (Aas et al., 1991). According to Babiak et al. (2001), the extender solution that worked well to cryopreserve spermatozoa of a species should posses the ability to maintain the sperm cell viability by inhibiting sperm motility. The key of success is via the use of an extender solution which is almost isotonic or mimicking the seminal plasma of that particular species. Therefore, it is crucial to know the seminal fluid osmolality of a species before media and diluents for that particular species can be formulated.

In all species of Malaysian fishes studied, sperm motility generally reduced after freezing and thawing process compared to the sperm motility before any freezing procedure. In P. jullieni, the motility of cryopreserved semen has reduced by approximately 45% compared to the fresh semen. A reduction of sperm motility by an average of 15% and 30% in T. tambroides and T. deuronensis respectively was observed. In general, sperm motility reduced between 10 - 70% on average in the species studied (Table 4). These observations are similar to studies in several other species such as Cyprinus carpio L. (Wamecke & Pluta, 2003), Oncorhynchus mykiss (Lahnsteiner et al., 1996) and Scophthalmus maximus (Dreanno et al., 1997).

Species	Motility %	
	Before freezing	Post-thaw
Javanese barb	88 - 97%	15 – 65%
African catfish	60 – 100%	15 – 70%
Isok barb	85 - 100%	30-84% (Mean: 49%)
Malaysian Mahseer	85 -100%	35-89% (Mean: 55%)
Local pangasiid catfish (Patin Buah)	85 -100%	35-70%
Kerai Kunyit	90 - 100%	35-80%
Tropical bagrid catfish (Muchlicin et al. 2004)	80 – 94%	58%

Table 4. Sperm motility (%) of the freshwater fish species studied before freezing and after thawing procedures

For all Malaysian fish species studied, the post-thaw motility rates of the cryopreserved sperm demonstrated similar values (p>0.05) even after a year of cryostorage as long as the semen samples are submerged well in the liquid nitrogen and without disturbance during the storage period.

4.2 Fertilization ability and hatching rate

In both Malaysian Mahseer and Isok barb, egg fertilization ability and hatching rates were found significantly lower (p<0.05) by using cryopreserved sperm compared with those fertilized using fresh sperm. The speed of embryos development was similar among the fertilized eggs using both cryopreserved and fresh sperm. Besides that, no significant difference (p>0.05) was found in the egg fertilization percentages between newly cryo-preserved semen and semen samples after a year of cryo-storage in both species. The performance of egg fertilization and hatching by using cryopreserved semen in four species of freshwater fish species studied in our laboratory is shown in Table 5.

Species	Fertilization (%)	Hatching (%)
Javanese barb	12 – 100	5 – 75
African catfish	21 - 37	19 – 32
Isok barb	1.2 – 10	0.8 – 4.6
Malaysian Mahseer	20 – 55	20 – 53

Table 5. Fertilization and hatching performance using cryopreserved sperm in several freshwater fish species carried out at FRI Glami Lemi

Quantitatively the numbers of hatchings from the cryopreserved semen were low, but qualitatively the hatchlings are visually normal and physically active and healthy as those from the fresh semen. Fingerlings produced from the cryopreserved semen from all four species showed normal development compared with those produced from the fresh semen.

5. Dependence of egg fertilization on semen cryopreservation methods

For the sperm cryopreservation of the fish species such as *P. jullieni* and *Tor* spp., several factors will determine the successful of the egg fertilization. The contributing factors that brought to the success of fish hatching using cryopreserved semen include viability of the post-thaw cryopreserved semen, good quality eggs, handling of the sperm and egg during the fertilization trials and knowledge in culturing the targeted species.

5.1 Viability of the cryopreserved sperm

Semen collection for cryopreservation should be done during the peak of spawning seasons of the particular species because the quality and quantity of spermatozoa are the highest at that particular time. The cryopreserved sperm cells tend to deteriorate very quickly and loss their motilily within 10-30 min after thawing procedure. Therefore, in quantification of the quality of the post-thaw sperm, the post-thaw motility shall be evaluated soon after thawing procedure in order to obtain good results. Similarly in the fertilization trials using cryopreserved semen, the mixture of semen to the eggs should be performed as quickly as possible once the frozen semen is thawed. Semen samples from each male broodfish were cryopreserved separately in order to maintain its inherent variability.

Successful cryopreservation procedures can maintain high quality of cryopreserved semen and this is revealed in high post-thaw sperm motility percentages. Successful cryopreservation procedures balance between the formation of ice crystals within the cells against excessive dehydration, which damages cellular structures (Tiersch, 2000). Each step involved in the cryopreservation procedures that was discussed in section 3 is equally important to produce viable cryopreserved sperm.

5.2 Good quality eggs

Good quality egg is also essential for successful fertilization when cryopreserved semen is used to produce offsprings. Same as sperm quality, egg quality is also the best during the peak of spawning seasons of the particular species. Thus fertilization trial is best to perform during spawning seasons of the fish. This is especially true and applicable for the seasonal bred species. In *P. jullieni*, significant higher fertilization rate (80-95%) was observed during the peak of the spawning season compared to the initial (50-60%) and toward the end of breeding season (30-50%) of the species. Healthy broodfish produced good quality eggs. Good management practice in broodstocks maintenance and good diet (i.e. high protein diet) are the key factors that produced healthy brood stocks.

5.3 Handling of sperm and egg during the fertilization trials

Good control in the timing of egg stripping and thawing of cryopreserved semen is important while using cryopreserved semen to fertilize eggs. Both egg and sperm should be made available at the same time in order to produce high fertilization rates. Spermatozoa of freshwater fishes are usually activated by the hypotonicity of their surrounding media (Morisawa & Suzuki, 1980; Morisawa et al., 1983a; Morisawa et al., 1983b). Most of the time and in majority species, hatchery water is sufficient as the sperm activation media and used in the sperm-egg mixture during fertilization process with cryopreserved semen. However, the use of suitable medium other than hatchery water for sperm activation is sometimes critical in some species (Lahnsteiner et al., 2003). This is observed in *T. tambroides* where specially formulated medium produced significant higher fertilization rate compared to the use of hatchery water as the sperm activation medium (Chew et al., 2010b). The effects of several media on egg fertilization ability had also reported by Billard (1983) on rainbow trout. Fertilization technique used in the fertilization trials is also one of the contributing factors to the success. Our studies showed that fertilization and hatching ability were significantly higher by using dry method compared to wet method.

5.4 Knowledge in culturing of the targeted species

Knowledge on the reproductive biology, broodstock management and husbandry, larva rearing and nursery of the particular fish species to be studied is a prerequisite for successful fertilization using cryopreserved semen. The age of maturity, breeding season (for seasonal bred species), factors that trigger and promote gonad maturation such as the type of nutrition and water quality where the brood fish is maintained are important factors that guaranteed gonad maturation and good health of the broodfish. Besides, well established artificial spawning method of the targeted species is also essential and such knowledge could later help production of fry using cryopreserved semen.

6. Current status of fish sperm cryopreservation in Malaysia –the fish semen cryobank

The current development of fish sperm cryopreservation in Malaysia is mainly focused on the creation of a fish semen cryobank. Such a cryobank for fish semen plays many important roles, especially for conservation of fish stocks and improve aquaculture. The establishment of a fish semen cryobank requires procurement of equipment and facilities for cryopreservation, skill development of the technical operators, choice of species for semen cryobanking, identification of the source for sample collection, developing of suitable cryopreservation procedure for the targeted species, proper record keeping and also proper maintenance of the semen cryobank.

6.1 Fish semen cryopresevation for conservation

Cryopreservation technology provides long term *ex situ* conservation of indigenous and endangered species (Mongkonpunya et al., 2000; Tiersch et al., 2000). Cryopreserved semen could be stored indefinitely without deterioration provided proper maintenance and handling measures are well taken care of all time (Stoss & Donaldson, 1983; Armitage, 1987; Suquet et al., 2000). In Malaysia, fish sperm cryopreservation was employed to save the genetics of the endangered and threatened indigenous species such as *P. jullieni* and *Tor* spp.. Fish species for gene banking and research are chosen on the basis of their threatened status and potential for aquaculture. Species with vulnerable, critically endangered (CR) and endangered (EN) status are prioritized in a conservation programme (Table 6). Species with protandrous behaviour throughout their life cycle such as the *Tenualosa toli* (Terubok) is also protected (MACD, 1996; Blaber et al., 2001; Wong, 2001). Other candidates of indigenous fish species which should be prioritized for semen cryopreservation are the Bala Shark (*Balantiocheilos melanopterus*), the indigenous catfish (*Clarias macrocephalus*) and several *Betta* spp. (*B. chini, B. hipposideros, B. livida, B. tomi* and *B. persephone*).

Semen samples could be stored for years via cryobanking before being used, and thus cryopreserved sperm banks can serve as insurance policies against unforseen catastrophes. However, very strict standardized protocols should be developed so that results are not biased by experimental and treatment variability but only by the inherent variability of the species. Further, these practices should be complemented with habitat conservation procedures (Bart, 2002). As it takes a long time to restore degraded ecosystems, the preserved semen is stored while upgrading of the ecosystems is going on. Both *in situ* and *ex situ* conservation should be applied complimentarily for sustainable management of the indigenous fisheries resources (Harvey, 1998). For an instance in the conservation programme of *P. jullieni*, introduction of closing season (from February to April each year) and conservation zones (12.4 km) at the identified spawning ground of *P. jullieni* at Pahang River have been implemented as the *in situ* conservation approach for the species (Zulkafli et al., 2010).

Conservation of indigenous freshwater fishes is a priority of the Department of Fisheries, Malaysia. Resulting from the success in breeding both *P. jullieni* and *Tor* spp. using cryopreserved semen, cryobanking of the semen of these species was initiated in 2008. However, those samples were collected from the cultured stock. Therefore, effort to further enrich the sperm cryobank through collection from wild stocks for different varieties and populations in the country is the ultimate aim of the Department of Fisheries, Malaysia. At present, semen samples from 88 *Tor* spp., 43 *P. jullieni*, 8 *P. nasutus* and 14 *H. wetmorei* have been collected and cryopreserved in the sperm cryobank of FRI Glami Lemi (Table 7).

The use of cryopreserved semen could support conservation efforts through stock enhancement and repopulation in areas where the species have declined or disappeared. In the breeding and restocking programmes, attempts to save the wild populations have so far largely focused on captive breeding or spawning of wild broodstock and subsequent release of hatchery-reared offsprings into the wild. Hatchery production of fry will support stock enhancement. Consequently, this will hopefully eliminate the need to harvest seed stock from the wild and the translocation of non-indigenous species for such programme.

Species	Status
Balantiocheilos melanopterus (Bala shark)	EN
Betta chini (Chini mouthbrooder)	VU
Betta hipposideros	VU
Betta livida (Emerald-sport fighting fish)	EN
Betta persephone (Black fighting fish)	CR
Betta tomi (Pikehead)	VU
Discherodontus halei (Spot-fin barb)	CR
Eleotris melanosoma	CR
Encheloclarias curtisoma (Soft fin walking catfish)	CR
Encheloclarias kelioides (Soft fin walking catfish)	CR
Encheloclarias prolatus (Catfish)	VU
Parasphromerus harveyi (Harvey's licorice gouramy)	EN
Probarbus jullieni (Isok barb)	EN
Scleropages formosus (Asian Arowana)	EN
Silurus furness	DD
Redigobius bikalanus	CR
Phallostethus dunckeri	VU
Sundoreonectes tiomanensis (Tioman cave loach)	VU
Helicophagus waandersii	EN
Cyclocheilichthys enoplus	EN
Leptobarbus hoeveni (Mad barb)	EN

(Source: DOF, Malaysia; Chong et al., 2010)

Table 6. List of indigenous freshwater fish species in Malaysia with critically endangered (CR), endangered (EN), vulnerable (VU) or data deficient (DD) status.

Species	Number of fish	Volume
Tor spp.	88	150 mL
P. jullieni	43	350 mL
P. nasutus	8	18 mL
H. wetmorei	14	30 mL

Table 7. Current status of cryogenic fish sperm bank of FRI Glami Lemi, Malaysia (since 2008)

6.2 Fish semen cryopreservation for aquaculture

Fish gene banks offer vast potential benefits to hatcheries (Munkittrick & Moccia, 1984; Chao & Liao, 2001). It offers genetic variability to fish hatcheries around the world. The use of frozen semen in breeding programmes offers a means to further broaden the genetic base of the targeted species. Genetic improvement of broodstock or hatchery species for traits such as disease resistant, fast growth rate, salinity tolerance etc. could also make feasible with the establishment of the cryogenic sperm bank. The applications of sperm cryopreservation in aquaculture were also highlighted by Mongkonpunya et al. (2000). In the case of some species, males and females reach maturity over different periods of time, the cryopreserved semen could facilitate artificial fertilization and seed production (Tiersch, 2000). Besides, cryopreserved semen is easier to transport than live fish for culturing. This eliminates the stress to fish. The risk of transmitting diseases is also reduced by using cryopreserved semen.

On the other hand, the use of cryopreserved sperm also provides flexibility in breeding programmes, especially in producing hybrids with favourable characteristics for culturing such as higher viability, intensive growth rate, adaptive flexibility, early sexual maturation etc. (FAO, 1971). Hybridization in fish culture becomes feasible and more manageable with the utilization of cryopreserved semen. For example, in a breeding programme of *H. wetmorei* carried out in our laboratory, *H. wetmorei* was cross-bred with the Javanese barb, *B. gonionotus* successfully via the use of cryopreserved sperm and surrogate egg from *B. gonionotus*. Such breeding procedure was needed because mature female of *H. wetmorei* was not available during the induced breeding programme of the species. Therefore, it is no doubt that the use of cryopreserved sperm provides greater control in breeding programmes.

7. Challenges to the gene banking of fish gamete via cryopreservation

7.1 Procedure optimization

The principle and process behind semen cryopreservation sound rather simple, i.e. the storage of semen samples in ultra-low temperatures and liquid nitrogen (-196°C) is normally used. In avian species (such as chicken, fowl, turkey, goose and duck) and mammal livestock species (such as cattle, horse, boar, sheep and goat), protocols for their semen cryopreservation are well developed and established (Hammerstedt & Graham, 1992; Curry, 2000; Donoghue & Wishart, 2000; Woelders et al., 2003). In fish however, there are no standard protocols that are applicable to all species. Unlike terrestrial animals, it is more difficult to standardize the semen cryopreservation protocols in fish (Tiersch, 2000). This is because different fish species exhibited different responses to the same extenders and cryoprotectants. For an example, the protocol or diluents formulation which served optimal to *P. jullieni* may not be necessarily suitable and served optimum in other species. As such, developing the species-specific and reproducible sperm cryopreservation procedure is thus required for this purpose. These include the choice of the type of extender solution and cryoprotectant, the rates of freezing and thawing etc. In general, the optimization of cryopreservation protocol for a species involved a series of complex interactions among various factors in each step.

7.2 Maintenance and proper recording

It is less costly to maintain a semen cryobank in a long run compared to *in situ* conservation approach such as live genebank. However, some costs need to be allocated for replenishing and topping up of liquid nitrogen in the storage dewars on periodically basis. Obtaining adequate funding and financial support to maintain the semen cryobank is part of the challenges to the institutions owned a semen cryobank. The monitoring of the viability of cryopreserved semen samples is also required from time to time to ensure the viability of stored samples is maintained. Post-thaw sperm motility was assessed for each batch of semen samples a week after cryo-storage and prior to be used for egg fertilization. The viability of the cryopreserved sperm was also evaluated from time to time through egg fertilization tests. A database that is able to provide good records of storage and withdrawals of samples from the semen cryobank is also a long term challenge in genebanking via cryopreservation (Kincaid, 2000). Proper record is essential for ease of samples retrieval in future.

7.3 Technical limitations

There are technical limitations to use cryopreserved semen in fish breeding as it requires involvement of skilled personnels. Therefore, training of operators or technicians in the related discipline is seen required in the technology extension programme. The small volume sample in straws (0.25 mL and 0.5 mL) is sufficient to be used in egg fertilization in laboratory based experiments and in genetic improvement programme of the targeted species. However, it is less practical to use cryopreserved semen for mass production of fry in aquaculture. Adaptation of the current developed protocols for practical application is thus important. The use of bigger straws volume or cryovials (5 mL or 10 mL) should be considered instead.

7.4 Difficulties in obtaining semen samples from wild populations

Difficulties in getting wild stocks are the main constraints in fish semen cryobanking of the indigenous species with threatened or endangered status and those species with high market demand. These limitations have caused difficulties in obtaining the effective population size (N_e), which is very crucial in future restoration efforts for the species. High quality seed is essential to support aquaculture. For many species especially of those riverine species, which their induced breeding method is still not established, their source of seed supply is still depend on the wild caught stocks. In Malaysia, it is also difficult to obtain fry and broodstock from hatchery because the lack of well organised hatchery operation. Each hatchery tends to maintain their own breeders, which is always limited in numbers. As the consequences, this resulted in high inbreeding rates among the hatchery stocks.

In Malaysia, it is now increasingly difficult for fish breeders to locate and collect genetic materials from healthy or relatively undisturbed populations in the wild. The loss of genetic material in fish species can hinder the development of the aquaculture industries, especially fish farms and hatcheries. Many hatcheries often rely on too few breeders to reproduce, resulting in lower production, susceptibility to diseases and poor survival rates in the wild. As wild fish stocks disappear, it becomes even more difficult for hatcheries to find new breeders. At the same time breeding within small populations with limited genetic diversity results in inbreeding depression, i.e. genetic drift, producing small or stunted fish stocks. Therefore, fish genetic resources must be conserved and utilized sustainably because they are the key to maintaining the viability of cultured and natural fish populations. They enable species to adapt to environmental change and also provide the opportunity for genetic improvement programme in aquaculture.

It is observed that large species that breed later in life are more vulnerable to fishing and changes in the environment, particularly in terms of fragmentation of their normal habitats. Indeed, most of the world's largest freshwater fish are at risk according to the IUCN Red List, and over exploitation contributes in a number of these cases. The dragon fish, *Scleropages formosus* is a well known case of over-exploitation. Therefore, giant indigenous species such as freshwater siluroids (*Wallago leeri,*), cyprinids (*Tor* spp., *Probarbus jullieni*), pangasiids (*Helicophagus waandersii*) etc. can be promoted as 'flagship species' or ecosystem ambassadors. At the same time, in terms of preserving biodiversity, by reducing the negative impacts of the continued spread of exotic fish species in the aquatic environment, efforts need to develop indigenous species for use in aquaculture. Those indigenous species that showed good adaptation to pond environment, resistance to handling, possess high growth rate and ability to reach sexually maturity in captivity are worth to be considered and developed as aquaculture species. To achieve the goal, we must safeguard indigenous fish resources both quantitatively and qualitatively from now before it is too late.

8. Conclusion

An overview of the current status of semen cryopreservation of Malaysian freshwater fishes is presented in this chapter. The role of semen cryobank is also discussed. Obviously, semen cryopreservation offers potential applications in *ex situ* conservation and sustainable management of the fisheries genetic resources in Malaysia, especially for those species with rare, vulnerable, threatened or endangered status, those protandry species with sex-changing characteristic over their life time, and also the potential candidates and genetic improved strains for aquaculture development. The successes in semen cryopreservation are very much relied on factors such as having ample knowledge on the biology and reproductive biology of the particular species of interest, trained personnels in various aspects such as gamete cryopreservation, breeding methods, broodstock management and husbandry and larva rearing and nursing of the targeted species.

9. Acknowledgement

We would like to thank the Department of Fisheries Malaysia for supporting this research through a development grant and also thanks to all research staffs of Freshwater Fisheries Research Division, FRI Glami Lemi and the Aquaculture Extension Centre, Perlok who have kindly provided assistance during the project.

10. References

Aas, G.H.; Refstie, T. & Gjerde, B. (1991). Evaluation of milt quality of Atlantic salmon. Aquaculture, 95: 125-132.

Ahmad-Ashhar, O. (1992). Induced spawning of Ikan Temoleh *Probarbus jullieni* Sauvage using human chrionic gonadotropin and heteroplastic crude piscine extract. In: Proceedings of the National IRPA Seminar, Kuala Lumpur, 9-11 January 1992. pp. 243-244.

Ahmad-Ashhar, O. (1996). Breeding and seed production technology of the Jelawat (*Leptobarbus hoevenii*).

Ahmad-Ashhar, O. (1998). Pembiakan Dan Pengeluaran Benih Ikan Jelawat (*Leptobarbus hoevenii*) Bleeker. *Buku Panduan Bil 2/98*. 21pp.

Ahmad-Ashhar, O. & Haron, A. (1994). Pembiakan aruhan Ikan Temoleh *Probarbus jullieni* Sauvage menggunakan ekstrak pituitari dan hormon human chorionic gonadotrophin (H.C.G.). *Proc. Fish. Res. Conf., DOF, Mal. IV*: 253-256.

Allan, J.D., Abell, R., Hogan, Z., Revenga, C., Taylor, B.W., Welcomme, R.L. & Winemiller, K. (2005). Overfishing of Inland Waters. *BioScience* 55(12): 1041-1051.

Ambak, M.A., Ashraf, A.H. & Budin, S. (2007). Conservation of the Malaysian Mahseer in Nenggiri basin through community action.

Babiak, I., Glogowski, J., Goryczko, K., Dobosz, S., Kuzminski, H., Strzezek, J. & Demianowicz, W. (2001). Effect of extender composition and equilibration time on fertilization ability and enzymatic activity of rainbow trout cryopreserved spermatozoa. *Theriogenology* 56: 177–192.

Baird, I. G. (2006). *Probarbus jullieni* and *Probarbus labeamajor*: the management and conservation of two of the largest fish species in the Mekong River in southern Laos. *Aquatic Conservation: Marine and Freshwater Ecosystems* 16(5): 517-532.

Baluyut, E. (1983). Stocking and introduction of fish in lakes and reservoirs in the ASEAN region. *FAO Tech. Pap.* 236. 82 pp.

Bart, A. (2002). Conservation of fish genetic diversity: need for development of a cryogenic genebank in Bangladesh. In: Penman, D.J., Hussain, M.G., McAndrew, B.J. and Mazid, M.A. (eds.) Proceedings of a workshop on Genetic Management and Improvement Strateges for Exotic Carps in Asia, 12-14 February 2002. Dhaka, Bangladesh. Pp 107-110.

Billard, R. (1983). Effects of cleolomic and seminal fluids and various saline diluents on the fertilizing ability of spermatozoa in the Rainbow trout, *Salmo gairdneri. J. Reprod. Ferti.*, 68: 77-84.

Billard, R. (1986). Spermatogenesis and spermatology of some teleost fish species. *Reprod. Nutr. Dev.*, 2: 877-920.

Billard, R. & Cosson, M.P. (1992). Some problems related to the assessment of sperm motility in freshwater fish. *J. Exp. Zool.*, 261: 122-131.

Billard, R., Cosson, J., Crim, L.W. & Suquet, M. (1995). Sperm physiology and quality. In: *Brood stock management and egg and larval quality.* (Eds) Bromage, N. R. and Roberts, R.J. Blackwell Science, Oxford. pp. 25-52.

Blaber, S.J.M., Milton, D.A., Brewer, D.T. & Salini, J.P. (2001). The shads (*genus Tenualosa*)of tropical Asia: An overview of their biology, status and fisheries. In: *Proceedings of the International Terubok Conference Sarawak, Malaysia, 11 -12 September 2001, Sarawak.* pp. 9-17.

Butts, I.A.E., Trippel, E.A. & Litvak, M.K. (2009). The effect of sperm to egg ratio and gamete contact time on fertilization success in Atlantic cod *Gadus morhua* L. *Aquaculture* 286 (1-2):89 – 94.

Chao, N.H. & Liao, I.C. (2001). Cryopreservation of finfish and shellfish gametes and embryos. *Aquaculture* 197: 161-189.

Chew, P.C., Hassan, R., Rashid, Z. A. & Chuah, H.P. (2010a). The endangered *Probarbus jullieni* (Sauvage) - Current status on its sperm cryopreservation in Malaysia. *Journal of Applied Ichthyology* 26: 797-805.

Chew, P.C., Rashid, Z. A., Hassan, R., Asmuni, M. & Chuah, H.P. (2010b). Semen cryo-bank of the Malaysian Mahseer (*Tor* spp.) *Journal of Applied Ichthyology* 26: 726-731.

Chong, V.C., Lee, P.K.Y. & Lau, C.M. (2010). Diversity, extinction risk and conservation of Malaysian fishes. *Journal of Fish Biology* 76: 2009-2066.

Curry, M.R. (2000). Cyopreservation of semen from domestic livestock. *Reviews of Reproduction* 5: 46-52.

Dobriyal, A.K., Kumar, N., Bahuguna, A.K. & Singh, H.R. (2000). Breeding ecology of some cold water minor carps from Garhwal Himalayas. Coldwater Fish and Fisheries: 177-186. (Eds) Singh, H.R. and Lakra, W.S. Narenda Publishing House, New Delhi. 337p.

DOF. (n.d.). Endangered freshwater fish species in Malaysia under the IUCN Red List. In: 3rd January 2011. *Available from:* <http://www.dof.gov.my/c/document_library/get_file?uuid=d8ebcbc5-4d4a-4125-8290-5d5bd78b1a5c&groupId=172176>.

Donoghue, A.M. & Wishart, G.J. (2000). Storage of poultry semen. *Animal Reproduction Science* 62: 213-232.

Dreanno, C., Suquet, M., Quemener, L., Cosson, J., Fierville, F., Normant, Y. & Billard, R. (1997). Cryopreservation of turbot (*Scophthalmus maximus*) spermatozoa. *Theriogenology* 48: 589-603.

Dudgeon, D. (2002). Fisheries : pollution and habitat degradation in tropical Asian rivers. *Encyclopaedia of Global Environmental Change, Vol. III* (ed. I. Douglas), pp. 316–323. John Wiley & Sons, Chichester, U.K.

FAO. (1971). Seminar/Study Tour in the U.S.S.R. on Genetic Selection and Hybridization of Cultivated Fishes. 19 April – 29 May 1968. Lectures. Rep.FAO/UNDP(TA), (2926): 360 p.

Froese, R., & Pauly, D. (2003). (Eds.). *Fishbase.* In: *World Wide Web electronic publication,* 3rd January 2011. *Available from:* <http://www.fishbase.org>.

Gwo, J.C. (2000). Cryopreservation of sperm of some marine fishes. In Tiersch, T.R. and Mazik, P.M. (Eds.), *Cryopreservation in Aquatic Species.* World Aquaculture Society, Baton Rouge, L.A. pp. 138-160.

Gwo, J.C., Strawn, K., Longnecker, M.T. & Arnold, C.R. (1991). Cryopreservation of Atlantic croakker spermatozoa. *Aquaculture* 94: 355-375.

Hammerstedt, R.H. & Graham, J.K. (1992). Cryopreservation of poultry sperm: the enigma of glycerol. *Cryobiology* 29: 26-38.

Harvey, B. (1998). An overview of action before extinction. In: Harvey B, Ross C, Greer D, Carolsfeld J (eds) Action before extinction: an international conference on conservation of fish genetic diversity. World Fisheries Trust, Victoria Inger, F.R. and Chin, P.K. 1962. *The Freshwater Fishes of North Borneo.* Chicago Natural History Muzeum, Chicago.

Ho, S.C. (1995). Status of Limnological Research and Training in Malaysia. In: Limnology in Developing Countries (eds B. Gopal and R.G. Wetzel). International Association for Limnology. International Scientific Publications, 50-B., Pocket C., Sidhartha Extension, New Delhi. Pp 163-189.

Hogan, Z., Baird, I.G. & Phanara, T. (2009). Threatened fishes of the world: *Probarbus jullieni* Sauvage, 1880 (Cypriniformes: Cyprinidae). *Environ. Biol. Fish* 84: 291-292.

Honeyfield, D.C. & Krise, W.F. (2000). Measuament of milt quality and factors affecting viability of fish spermatozoa. In: Tiersch, T.R. and Mazik, P.M. (Eds.), *Cryopreservation in Aquatic Species.* World Aquaculture Society, Baton Rouge, L.A. pp. 49-58.

Horvath, A., Miskolczi, E. & Urbányi, B. (2003). Cryopreservation of common carp sperm. *Aquat. Living Resour.* 16: 457–460.

Hung, L. T., Suhenda, N., Slembrouck, J., Lazard, J. & Moreau, Y. (2004). Comparison of dietary protein and energy utilization in three Asian catfishes (*Pangasius bocourti, P. hypophthalmus* and *P. djambal*). *Aquaculture Nutrition* 10: 317-326.

Inger, R.F. (1955). Ecological notes on the fish fauna of a coastal drainage of north Borneo. *Fieldiana Zool.* 37: 47 -90.

Inger, R.F. & Chin, P.K. (1962). The Freshwater Fishes of North Borneo. *Fieldiana Zool.* 45: 1-268.

Inger, R.F. & Chin, P.K. (2002). *The Freshwater Fishes of North Borneo with a supplementary chapter by Chin Phui Kong.* Natural History Publications (Borneo), Kota Kinabalu. 268 pp. + S-78.

Ingram, B., Sungan, S., Gooley, G., Sim, S.Y., Tinggi D. & de Silva S.S. (2005). Induced spawning, larval development and rearing of two indigenous Malaysian Mahseer, *Tor tambroides* and *T. duoronensis. Aquacult. Res.* 36(10): 1001-1014.

IUCN (2009). IUCN Red List of Threatened Species. Version 2009.1. <www.iucnredlist.org>. Downloaded on 25 August 2009.

Jackson, D.C. & Marmulla, G. (2001). The influence of dams on river fisheries. In Dams, fish and fisheries: Opportunities, challenges and conflict resolution (ed. Marmulla, G.). *FAO Fisheries Technical Paper* 419: 1-45.

Jantrarotai, P., Sutthiwises, S., Kamonrat, V., Peyachoknagul, S. & Vidthayanon, C. (2007). Species identification of 3 *Hypsibarbus* spp. (Pisces: Cyprinidae) using PCR-RFLP of cytochrome b gene. *Kasetsart J. (nat. Sci.)* 4: 660 – 666.

Khan, M.S. (1987). Some aspects of the biology of ikan baung, *Mystus nemurus* C. & V. with reference to Chenderoh Reservoir. Thesis of MSc. Universiti Pertanian Malaysia, Serdang.

Khan, M.S., Ambak, M.A., Ang, K.J. & Mohsin, A.K.M. (1995). Reproductive biology of a tropical catfish, *Mystus nemurus* Cuvier and Valenciennes, in Chenderoh Reservoir Malaysia. *Aquaculture and Fisheries Management* 21: 173-179.

Khoo, K.H., Leong, T.S., Soon, F.L., Tan, S.P. & Wong, S.Y. (1987). Riverine fisheries in Malaysia. *Archiv. feur Hydrobiologie Beiheft.* 28: 261 – 268.

Kincaid, H.L. (2000). Development of Databases for Germplasm Repositories. In: Tiersch, T.R. and Mazik, P.M. (eds) *Cryopreservation in Aquatic Species.* World Aquaculture Society, Baton Rouge.Pp 323-331.

Kopeika, E., Kopeika, J. & Zhang, T. (2007). Cryopreservation of fish sperm. In: Day, J.G. and Stacey, G.N. (Eds) *Methods Mol. Biol. Vol.* 368: *Cryopreservation and Freeze-Drying Protocols, 2nd edition.* Humana Press Inc., Totowa, NJ. pp. 203 -217.

Kottelat, M., Whitten, A.J., Kartokasari, S.N. & Wirjoratmodjo, S. (1993). *Freshwater Fishes of Western Indonesia and Sulawesi.* Singapore: Berkeley Book Pte Ltd.

Kurokura, H., Hirano, R., Tomita, M. & Iwahashi, M. (1984). Cryopreservation of carp sperm. *Aquaculture* 37: 267-273.

Lahnsteiner, F. & Patzner, R.A. (2008). Sperm morphology and ultrastructure in fish. In: Alavi, S.M.H., Cosson, J.J., Coward, K. and Rafiee, G. (Eds) *Fish Spermatology.* Alpha Science International Ltd., Oxford, UK. pp. 1-61.

Lahnsteiner, F., Patzner, R.A. & Weismann, T. (1996). Semen cryopreservation of salmonid fishes: influence of handling parameters on the post thaw fertilization rate, Aquacult. Res. 27(9): 659–671.

Lahnsteiner, F., Weismann, T. & Patzner, A.R. (1997). Methanol as cryoprotectant and the suitability of 1.2 ml and 5 ml straws for cryopreservation of semen from salmonid fishes. *Aquaculture Research* 28: 471-479.

Lahnsteiner, F., Berger, B. & Weismann, T. (2003). Effects of media, fertilization technique, extender, straw volume, and sperm to egg ratio on hatchability of cyprinid embryos, using cryopreserved semen. *Theriogenology* 60: 829-841.

Leung, L.K.P. & Jamieson, B.G.M. (1991).Live preservation of fish gametes. In: Jamieson, B.G.M. (Ed.), *Fish Evolution and Systematics: Evidence from Spermatozoa*. Cambridge Univ. Press, Cambridge, pp. 245–269.

Maack, G.G. & George, M.R. (1999). Contributions to the reproductive biology of *Encrasicholina punctifer* Fowler, 1938 (engraulidae) from West Sumatra, Indonesia. *Fisheries Research* 44: 113 – 120.

MACD (1996). Life Cyle of Terubok (*Tenalosa toli*) With Recommendations for the Conservation, Management and Culture of The Species. Phase 3 Final Report. A collaborative project of CSIRO Division of Fisheries, Marine Laboratories, Cleveland Queensland, Australia and Inland Fisheries Branch, Ministry of Agriculture & Community Development (MACD), Kuching. Pp.64.

Mohamad-Zaini S. (1992). Pembiakan Aruhan Dan Pengeluaran Benih Ikan Sebarau *Hampala macrolepidota*. *Buletin Perikanan* 82. 13 pp.

Mohd-Zafri, H. (2006). Morphology and general reproductive stages of *Pangasius nasutus* from Sg. Pahang in Maran District, Pahang, Malaysia. Thesis Msc, UPM.

Mohsin, A.K.M. & Ambak, M.A. (1991). *Freshwater Fishes of Peninsular Malaysia*. Penerbit Universiti Pertanian Malaysia, Serdang, Selangor.

Mongkonpunya, K., Pupipat, T. & Tiersch, T.R. (2000). Cryopreservation of sperm of Asian Catfishes including the endangered Mekong Giant Catfish. In: Tiersch, T.R. and Mazik, P.M. (Eds.), *Cryopreservation in Aquatic Species*. World Aquaculture Society, Baton Rouge, L.A. pp. 108-116.

Morisawa, M. & Suzuki, K. (1980). Osmolality and potassium ion: Their roles in initiation of sperm motility in eleosts. *Science* 210: 1145 – 1147.

Morisawa, M., Suzuki, K. & Morisawa, S. (1983a). Effects of potassium and osmolality on spermatozoan motility of Salmonid fishes. *Journal of Experimental Biology* 107: 105 – 113.

Morisawa, M., Suzuki, K., Shimizu, H., Morisawa, S. & Yasuda, K. (1983b). Effects of potassium and osmolality on motility of spermatozoa from freshwater cyprinid fishes. *Journal of Experimental Biology* 107: 95 – 103.

Muchlisin, Z.A., Hashim, R. & Chong, A.S.C. (2004). Preliminary study on the cryopreservation of tropical bagrid catfish (*Mystus nemurus*) spermatozoa: the effect of extender and cryoprotectant on the motility after short-term storage. *Theriogenology* 62: 25-34.

Munkittrick, K.R. & Moccia, R.D. (1984). Advances in the cryopreservation of salmonid semen and suitability for a production-scale artificial fertilization program. *Theriogenology* 21: 645-59.

Ng, C.K. (2004). *Kings of the Rivers – Mahseer in Malaysia and the region*. Inter Sea Fishery (M) Sdn. Bhd. Selangor.

Nguyen, T.T.T. (2008). Population structure in the highly fragmented range of *Tor douronensis* (Cyprinidae) in Sarawak, Malaysia revealed by microsatellite DNA markers. *Freshwater Biology* 53: 924-934.

Nguyen, T.T.T., Baranski, M., Rourke, M. & McPartlan, H. (2006). Characterisation of microsatellite DNA markers for the mahseer species, *Tor tambroides* and cross-amplification in other four congeners (*T. douronensis, T. khudree, T. putitora*, and *T. tor*). *Molecular Ecology Notes* 7: 109-112.

Nguyen, T.T.T., Ingram, B., Sungan, S., Gooley, G., Sim, S.Y., Tinggi, D. & De Silva, S.S. (2006). Mitochondrial diversity of broodstock of two indigenous mahseer species, *Tor tambroides* and *T. douronensis* (Cyprinidae) cultured in Sarawak, Malaysia. *Aquaculture* 253: 259-269.

Nguyen, T.T.T., Ingram, B., Sungan, S., Tinggi, D., Sim, S.Y., Gooley, G. & De Silva, S. S. (2007). Guidelines for genetic management and conservation of two mahseer species in Sarawak, Malaysia. (Downloaded on 20 October 2009)

Nguyen, T.T.T., Sukmanomon, S., Na-Nakorn, U. & Ziming, C. (2008). Molecular phylogeny and phylogeography of mahseer species (Cyprinidae: Cypriniformes) as inferred from sequences of two mitochondrial gene regions. *Molecular Phylogenetics and Evolution* (in press) (doi:10.1016/j.ympev.2008.01.006).

Pathmasothy, S. (1985). The Effect of Three Diets with Variable Protein Level on Ovary Development and Fecundity in *Leptobarbus hoevenii*. In: Cho, C.Y., D.S. Cowey and W. Watanabe, Finfish Nutrition in Asia; Methodological Approach to Research and Development. Proceedings of the Asian Finfish Nutrition Workshop Held in Singapore 23-2 August, 1983. Ottawa: IDRC., pp.107-112.

Pathmasothy, S., & R. Omar. (1982). The Effect of Four Different Diets on the Growth of *Leptobarbus hoevenii*. *MARDI Research Bulletin* 10(1): 100-113.

Rainboth, W.J. (1996). *Fishes of the Cambodian Mekong. FAO Species Identification Field Guide for Fishery Purposes.* Food and Agriculture Organization (FAO) Publication, Rome.

Rideout, R.M., Litvak, M.K. & Trippel, E.A. (2003). The development of a sperm cryopreservation protocol for winter flounder, *Pseudopie-uronectes americanus* (Walbaum): evaluation of cryoprotectants and diluents. *Aquaculture Research* 34: 653-659.

Roberts, T.R. (1989). *The Freshwater Fishes of Western Borneo (Kalimantan Barat, Indonesia).*California Academy of Sciences, California.

Roberts, T.R. (1992). Revision of the Southeast Asian cyprinid fish genus *Probarbus*, with two new species threatened by the proposed construction of dams on the Mekong River. *Ichthyl. Expl. Freshw.* 3: 37-48.

Roberts, T.R. & Vidthayanon, C. (1991). Systematic revision of the Asian catfish family Pangasiidae, with biological observations and description of three new species. *Proceedings of the Academy of NaturalSciences of Philadelphia* 143: 97-144.

Saidin, T. (1986). Induced spawning of *Clarias macrocephalus* (Gunther). In: Maclean, J.L., Dizon, L.B. & Hosillos, L.V. (eds.) *The First Asian Fisheries Forum.* Asian Fisheries Society, Manila, Philipines. P. 683 – 686.

Scott, A.P. & Baynes, S.M. (1980). A review of the biology, handling and storage of salmonid spermatozoa, J. Fish Biol. 17: 707-739.

Stoss, J & Donaldson, E.M. (1983). Studies on cryopreservation of eggs from rainbow trout (*Salmo gairdneri*) and coho salmon (*Oncorhynchus kisutch*). *Aquaculture* 31: 51 – 65.

Suhairi, A. (1996). Preliminary observations on the culture of the Temoleh (*Probarbus jullieni*) in ponds. Paper presented in *Fisheries Semeniar July 1996,* IPP Batu Maung, Pulau Pinang, Malaysia.

Suhairi, A., Misri, S., Abdul Ghani, H. & Ahmad Ashhar, O. (1996). Kajian awal ternakan ikan temoleh (*Probarbus jullieni*) dalam kolam. *Proc. Fish. Res. Conf., DOF, Mal., 1996*: 278-283.

Suquet, M., Omnes, M.H., Normant, Y. & Fauvel, C. (1992). Assessment of sperm concentration and motility in turbot *Scophthalmus maximus*. *Aquaculture* 101: 177-185.

Suquet, M., Dreanno, C., Fauvel, C., Cosson, J. & Billard, R. (2000). Cryopreservation of sperm in marine fish. *Aquaculture Research* 31: 231-243.

Thalathiah, S., Abas Fauzi, O. & Ibrahim, T. (1988). First successful attempt to induce breed *Mystus nemurus* (C & V) at Batu Berendam, Melaka. *Proc. 11th Ann. Conf. MSAP:* 53 – 55.

Thalathiah, S., Ibrahim, T. & Mansor, A. (1992). Induced spawning of *Mystus nemurus* (C & V) using heteroplastic pituitary extract, HCG and an analog of LHRH. In: *Proceedings of Fisheries Research Seminar, Malacca, 27 – 29 June 1989. Pp. 185 – 188.*

Tiersch, T.R. (2000). Introduction. In: Tiersch, T.R. and Mazik, P.M. (Eds.), *Cryopreservation in Aquatic Species.* World Aquaculture Society, Baton Rouge, L.A. pp. 19-26.

Tiersch, T.R. (2006). Fish sperm cryopreservation for genetic improvement and conservation in Southeast Asia. *Fish For The People* 4(2): 21-33.

Tiersch, T.R., Goudie, C.A. & Carmichael, G.J. (1994). Cryopreservation of channel catfish sperm: storage in cryoprotectants, fertilization trials, and growth of channel catfish produced with cryopreserved sperm. *Transactions of the American Fisheries Society* 123: 580-586.

Tiersch, T.R., Figiel, C.R., Wayman, Jr., W.R., Williamson, J.H., Carmichael, G.J. & Gorman, O.T. (2000). Cryopreservation of sperm of the endangered razorback sucker. In: Tiersch, T.R. and Mazik, P.M. (Eds.), *Cryopreservation in Aquatic Species.* World Aquaculture Society, Baton Rouge, L.A. pp. 117-122.

Tsigenopoulos, C.S. & Berrebi, P. (2000). Molecular phylogeny of North Mediterranean freshwater barbs (Genus Barbus: Cyprinidae) inferred from cytochrome b sequences: biogeographic and systematic implications. *Molecular Phylogenetics and Evolution* 14: 165-179.

Wamecke, D. & Pluta, H.J. (2003). Motility and fertilizing capacity of frozen/thawed common carp (*Cyprinus carpio* L.) sperm using dimethyl-acetamide as the main cryoprotectant. *Aquaculture* 215: 167-185.

Woelders, H. Zuidberg, C.A. & Hiemstra, S.J. (2003). Applications, limitations, possible improvements and future of cryopreservation for livestock species. In: *Proceedings of the Workshop on Cryopreservation of Animal Genetic Resources in Europe. Paris, 23rd February 2003.* pp 67-76.

Wolf, K. (1963). Physiological salines for freshwater teleosts. *The Progressive. Fish Culturist.* 25: 135-140.

Wong, P. P. H. (2001). The status of the Terubok (*Tenualosa toli*) fishery in Sarawak, Malaysia. In: *Proceedings of the International Terubok Conference Sarawak, Malaysia, 11 -12 September 2001, Sarawak.* pp. 91-99.

Zhang, Y.Z., Zhang, S.C., Liu, X.Z., Xu, Y.Y., Wang, C.L., Sawant, M.S., Li, J. & Chen, S.L. (2003). Cryopreservation of flounder (*Paralichthys olivaceus*) sperm with a practical methodology. *Theriogenology* 60: 989-996.

Zhou, W. & Chu, G.H. (1996). A review of *Tor* species from the Lancangjiang River (Upper Mekong River), China (Teleostei: Cyprinidae). *Ichthyological Exploration of Freshwater* 7: 131-142.

Zulkafli A.R., Chew, P.C. & Johari, I. (2010). Conservation of Freshwater Fishes and Enhancement Programmes in Peninsular Malaysia. 2nd *National Conference on Agrobiodiversity Conservation and Sustainable Utilization (NAC2)-Agrobiodiversity for sustainable economic development, 11-13 May 2010, Tawau, Sabah. Pp 48-50.*

2

Marine Fish Sperm Cryopreservation and Quality Evaluation in Sperm Structure and Function

Qing Hua Liu, Zhi Zhong Xiao, Shi Hong Xu,
Dao Yuan Ma, Yong Shuang Xiao and Jun Li
Center of Biotechnology R&D, Institute of Oceanology,
Chinese Academy of Sciences, Qingdao
PR China

1. Introduction

Long-term storage of sperm in liquid nitrogen is a valuable technique for genetic resources preservation (Kopeika et al. 2007). The research on fish sperm cryopreservation has achieved great advances since the first successful sperm cryopreservation in herring (Blaxter 1953). It provides many benefits such as ease of global germplasm shipping and supply (Tiersch et al. 2004), selective breeding and hybridization with desirable characteristics (Henderson-Arzapalo et al. 1994), and conservation of genetic diversity (Van der Walt et al. 1993; Tiersch et al. 2000; Ohta et al. 2001). Furthermore, a frozen sperm bank could maintain the continuous and stable supply of gametes for hatchery seed production or laboratory experimentation. Because of the advantages of this technique, fish sperm of over 200 freshwater and 40 marine species have been cryopreserved successfully (Gwo 2000).

Most of fish sperm cryopreservation researches have focused on freshwater species such as cyprinids (Babiak et al. 1997; Lahnsteiner et al. 2000), salmonoids (Conget et al. 1996; Cabrita et al. 2001), catfishes (Christensen and Tiersch 1997; Viveiros et al. 2000) and loach (Kopeika 2003a, b; Dzuba & Kopeika 2002). In recent years, with the rapid development of marine fish aquaculture, some experiments on germplasm cryopreservation have also been conducted in marine fish species, especially the great commercial value ones such as red seabream (Liu, et al. 2006 ; Liu, et al. 2007a，b ; Liu, et al. 2010 a，b) turbot (Dréanno et al. 1997; Chen et al. 2004), flounder (Richardson et al. 1999; Zhang et al. 2003), and halibut (Billard et al. 1993).

Damage to sperm morphology and function usually occurs during the process of freezing and thawing. Cellular damage may greatly decrease motility, impair velocity, and reduce fertilizing capacity, even lead to DNA strand breakage or mutation (Dréanno et al, 1997; Lahnsteiner et al, 1996a; Warnecke & Pluta 2003; Kopeika et al, 2004). Although motility and fertilizing capacity are usually assessed in frozen-thawed sperm, these methods have limitations. Many factors affect the validity of these assessments, including subjectivity, microscope performance, the quality of eggs, and fertilization protocols. Some new technologies have been used in fish sperm quality analysis, such as computer-assisted sperm analysis (CASA), being used to objectively evaluate sperm motility (Lahnsteiner et al., 1996b; Lahnsteiner et al., 1998; Kime et al., 1996) ; Electron microscopy, being used to detect cryodamage (ultrastructural changes) in frozen-thawed sperm (Zhang et al, 2003, He &Woods 2004); In addition, flow cytometry of fluorescent-stained sperm have been used in mammals (Graham et al, 1990; Gravance et al, 2001) and turkeys (Donoghue et al ,1995), providing rapid, precise information regarding the viability of thousands of individual sperm. In recent years, flow cytometry has also been successfully used to assess both fresh and cryopreserved fish sperm (Ogier de Baulny et al, 1999; Segovia et al, 2000).

Red seabream is one of the most commercially important marine fish species for aquaculture in China. However, the decline of wild red seabream population has occurred due to over fishing and marine pollution in recent years. The use of cryopreserved sperm can provide an efficient method to increase its genetic population size and to help maintain genetic diversity. The aims of this study were to establish efficient methods for cryopreservation of red seabream sperm with 2-mL cryovials and to objectively measure the post-thaw sperm motility characteristics by means of CASA, to evaluate the post-thaw sperm fertilization capacity, and the cryodamage by electron microscopy and flow cytometry.

2. Sperm cryopreservation and quality evaluation

2.1 Materials and methods

2.1.1 Gametes collection

Naturally matured fishes were obtained from Qingdao hatchery during the spawning season (From the middle of March to the end of May). Twenty males and 10 females (3 kg to 4 kg individually, 10 years old) were cultivated in a 20-m3 concrete rearing pond with flow-through seawater and fed daily with cooked meat of bay mussel, *Mytilus edulis*. Prior to handling, males were firstly anesthetized in a 0.003% eugenol bath. Sperm was collected into petri dishes by gently hand-stripping the abdomen of the ripe males. Extreme care was taken to avoid the contamination of sperm with seawater, blood, urine and feces. The percentage of motile spermatozoa was checked with a Nikon-YS-100 light microscope (Nikon Corporation, Tokyo, Japan) at 250 × magnification. Sperm with motility > 85% was kept on crushed ice and transported to the laboratory for further use. Eggs were collected by abdominal pressure of the females at the time of ovulation. Good eggs were slightly yellowish, translucent and round-shaped. Eggs for fertilization trials were collected only from one female.

2.1.2 General procedure for sperm freezing and thawing

Sperm were diluted in Cortland extenders (Liu et al, 2006) containing DMSO with different concentrations (6–24% DMSO). After mixing thoroughly, 1.6 ml sperm was placed into 2-ml cryovials. The cryovials were transferred into a Kryo-360-1.7 programmable freezer (Planer Plc. Middlesex, UK), equilibrated for five minutes at 0oC, and frozen from 0 to −150oC at a cooling rate of 20oC min-1, then plunged into liquid nitrogen for storage. The frozen sperm were thawed in 40oC water bath after being preserved in liquid nitrogen for one month. After that, the thawed sperm was evaluated for motility and fertilizing capacity.

2.1.3 Sperm motion characteristics analysis by using CASA

Sperm motion characteristics were assessed by using a computer-assisted sperm motion analysis system (CASAS-QH-III, Tsinghua Tongfang Inc., Beijing, China) at room temperature (18°C to 20°C). The method for computer-assisted sperm motion analysis was described in Liu et al (2007b). The designation of the motility status was based on the level of the average path velocity (VAP). Sperm with average path velocity <5 µm s[-1] were considered immotile, with average path velocity >20 µm s[-1] were defined as motile, and 5–20 µm s[-1] as locally motile. Therefore, in the present study sperm motility includes the percentage of local motile sperm and motile sperm. Motility and velocity of fresh and post-thaw sperm were quantitatively recorded by CASA immediately 10 s after activation, and changes of motility of post-thaw sperm frozen with 15% DMSO were observed every 30 s.

2.1.4 Sperm fertilization and hatching experiments

Fertilization capacity of post-thaw sperm frozen with DMSO (6–24% DMSO) was evaluated. The optimized sperm to egg ratio of 500:1 was selected for the following fertilization trials (Li et al., 2006). The artificial fertilization method was described in detail in Liu et al (2007b). Fertilization rates were evaluated within 6–8 h after insemination by counting the percentage of gastrula-stage embryos in relation to the total number of eggs used. Forty-eight hours after fertilization, the number of hatched larvae was counted in each experiment. The hatching rates were calculated as the percentage of hatched larvae in relation to the total number of eggs used in each experiment.

2.1.5 Ultrastructure

Prior to scanning electron microscopy, sperm were fixed in 2.5% glutaraldehyde diluted in PBS (pH 7.6), dehydrated in a series of increasing concentrations of ethanol, critical-point dried, evaporated with gold, and examined with a scanning electron microscope (KYKY-2800B; KYKY Technology Development Ltd., Beijing, China) For transmission electron microscopy, spermwas prefixed in 2.5% glutaraldehyde, post-fixed in 1% osmium tetroxide, and embedded in Epon 812. Ultrathin sections were prepared, counterstained with 2%uranyl acetate followed by lead citrate, and examined with a transmission electron microscope (HITACHI H- 7000; Hitachi Ltd., Tokyo, Japan), and the number of sperm with various categories (normal, slightly damaged, and seriously damaged sperm) of cryodamage was determined. One-hundred sperm were randomly selected for observation each time; this was repeated three times on different sections (total of 300 frozen-thawed sperm for each male).

2.1.6 Rhodamine 123, propidium iodide and flow cytometry

The staining method used was described in (Liu et al. 2007a). An aliquot of mixed fresh or frozen-thawed sperm with 15%DMSO was incubated for 20 min (in the dark, temperature 4 °C) with 5 mg/mL of Rhodamine 123 (Rh123, Sigma Chemical Co., St. Louis, MO, USA).

Thereafter, sperm were incubated for 45 min in 1.5 mL of Cortland extender. With this staining method, only cells with functional mitochondria were stained, due to the negative potential of the inner membrane of the mitochondria. Samples were diluted and counterstained with 5 mg/mL of propidium iodide (PI, Sigma Chemical Co.). After 10 min, sperm samples were analyzed with flow cytometry (FACSvantage SE flow cytometer; Becton Dickinson, Mountain View, CA, USA) as previously described for trout sperm (Ogier de Baulny, et al, 1997). Sperm populations were identified according to their relative red and green fluorescence (staining with PI and Rh123, respectively). Sperm with red (stained with PI) DNA were interpreted as having a damaged plasma membrane, whereas those that were green (stained with Rh123) were interpreted as having intact mitochondrial function. Sperm that were only red (damaged plasma membrane and lacking mitochondrial function), were localized in Region 1, whereas those that were only green (intact membrane and functional mitochondria), were localized in Region 3. Sperm with both red and green fluorescence (damaged plasma membrane and functional mitochondria) were localized in Region 4, and those with no staining (intact plasma membrane, but no mitochondrial activity) were localized in Region 2.

2.1.7 Statistical analysis

To determine the effects of cryopreservation on sperm motility, fertilization capacity, structure and function, a paired-sample t-test was used to compare fresh versus frozen-thawed sperm. All statistical analyses were performed with SPSS Version 11.0 software (SPSS Inc. Chicago, IL, USA) and $P < 0.05$ was considered significant. All data were expressed as mean±S.D.

2.2 Results

2.2.1 Post-thaw sperm viability

The influence of cryopreservation on sperm motility and velocity was shown in Table 1. Percentages of motile post-thaw sperm frozen with 12–21% DMSO were higher than those with 6% DMSO, 9% DMSO and 24% DMSO. However, the procedure of cryopreservation has no significant ($P>0.05$) influence on the motile sperm velocity 10 s after activation compared with fresh sperm. In addition, the post-thaw sperm frozen with 12-21% DMSO showed similar types of straight trajectories.

Cryoprotectant (%)	Motility parameters		
	Locally motile (%)	Motile (%)	Velocity (μm s^{-1})
Fresh sperm	22.0±7.7	64.7±14.2 c	113.1±10.6 a
6% DMSO	16.6±4.6	26.8±11.4 a	89.1±15.0 a
9% DMSO	21.3±7.2	40.3±9.1 ab	91.9±13.5 a
12% DMSO	17.8±10.6	61.6±8.5 c	95.2±12.3 a
15% DMSO	20.4±6.1	64.8±8.7 c	99.3±11.6 a
18% DMSO	21.8±3.9	62.9±6.2 c	97.7±15.2 a
21% DMSO	16.4±4.3	60.8±5.4 c	90.1±12.3 a
24% DMSO	16.6±7.9	55.7±9.2 bc	95.7±8.9 a

Table 1. The influence of cryopreservation on sperm motility and velocity in *P. major*

The motion characteristics of fresh and post-thaw sperm were evaluated by using computer-assisted sperm analysis 10 s after activation. This table shows the percentages of locally motile (VAP range from 5 to 20μm s-1) and motile sperm (VAP > 20μm s-1) as well as their velocity (VAP) for fresh and post-thaw sperm. Values superscripted by the same letter are not significantly different ($P>0.05$, n=5).

The effect of time after activations on post-thaw sperm motility was shown in Fig. 1. The percentages of total motile sperm of both fresh (87. 2 ± 6.1%) and post-thaw sperm (81.9 ± 6.6%) frozen with 15% DMSO were not ($P>0.05$) different significantly 10 s after activation. However, 30 s after activation the percentage of total motile post-thaw sperm (72.3 ± 6.3%) was ($P<0.05$) lower than that of fresh sperm (82.7 ± 7.2%). Sixty seconds after activation, the percentage of post-thaw sperm motility drastically reduced to 38.7 ± 13.2%.

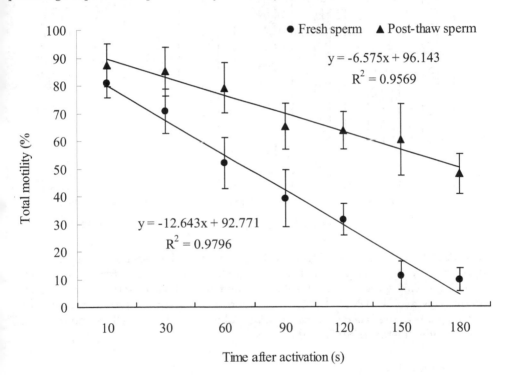

Fig. 1. The influence of time after activation on the motility of fresh and post-thaw sperm in P. major. Ten seconds after activation, the total motilities of fresh and post-thaw sperm frozen with 15% DMSO were observed every 30 s using computer-assisted sperm analysis system. This figure describes the evolution of the total motilities of fresh (▲) and post-thaw sperm (●) after activation respectively (n =5).

2.2.2 Post-thaw sperm fertilizing capacity and hatchability

Fertilization rates and hatching rates of fresh and post-thaw sperm were shown in Fig. 2. The fertilization rates and hatching rates were similar for fresh and post-thaw sperm frozen with 12–21% DMSO. However, lower ($P<0.05$) fertilizing capacity of post-thaw sperm frozen with 6% DMSO, 9% DMSO and 24% DMSO were observed. In addition, the percentages of motile of post-thaw sperm and fertilization rates showed a high positive linear regression (r = 0.876). Similarly, the percentages of motile spermatozoa and hatching rates of post-thaw sperm showed a high positive linear regression (r = 0.878).

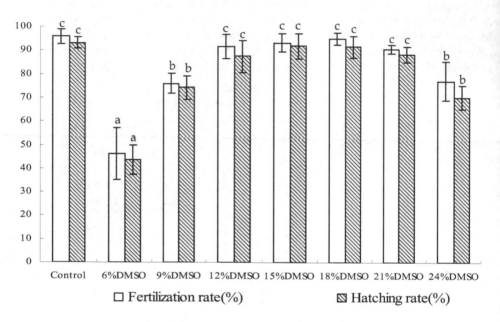

Fig. 2. Fertilization rates and hatching rates of fresh and post-thaw sperm in P. major. Cryopreserved sperm was thawed and activated for the artificial fertilization with sperm to egg ratio 500:1. This figure describes the fertilization rates and hatching rates of post-thaw sperm frozen with 6-24% DMSO. □ For fertilization rates of fresh and post-thaw sperm; ▨ For hatching rates of fresh and post-thaw sperm. Columns marked with the same letter are not significantly different ($P>0.05$, n=5).

2.2.3 Sperm ultrastructure

Ultrastructure of fresh and intact frozen-thawed red seabream sperm are shown in Fig. 1. These sperm had a head, midpiece, and tail. The head was ovoid and contained the nucleus and centriolar complex; the latter consisted of two centrioles. The midpiece was approximately cylindrical and contained mitochondria. The flagellum consisted of nine peripheral doublets and two central microtubules; the axoneme was a typical 9 + 2 structure (Fig. 3 A, B). The proportion of fresh sperm with normal morphology was 77.8 ± 5.6%, whereas after cryopreservation, 63.0 ± 7.2% of the sperm had normal morphology (Fig. 3 C), 20.7 ± 3.1% were partly damaged (e.g. swelling or rupture of head, midpiece and tail region, as shown in Fig. 3 D, as well as damage to mitochondria). Furthermore, 16.4 ± 4.2% were severely damaged; the plasma membranes was completely ruptured and only nuclei, mitochondria, or some fragments of cellular organelles were found (Fig.3 E).

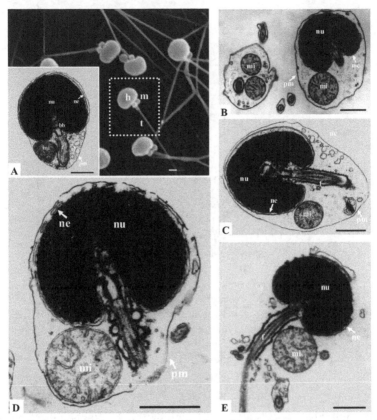

Fig. 3. The morphology and ultrastructure of fresh and normal post-thaw spermatozoa of red seabream. (A) Total view of fresh spermatozoa and the internal structure of head and mid-piece of fresh spermatozoa. (B) Flagellum of fresh spermatozoa. (C) Unchanged spermatozoa cryopreserved with 15% DMSO. (D) Partly damaged spermatozoa. (E) Completely damaged plasmalemma and nuclear envelop. (h, head; m , mid-piece; t, tail region. nu, nucleus; ne, nuclear envelope; bb, basal body; mi, mitochondrion; pm, plasmalemma; f, flagellum; v, vacuole). Scale bar = 0.5 µm.

2.2.4 Fluorescent staining and flow cytometry

Sperm populations were localized into four distinct regions according to their relative green and red fluorescence after staining with PI and Rh123 (Fig. 4). For fresh sperm, 83.9% had an intact membrane and functional mitochondria, 5.1% had nonfunctional mitochondria, 9.8% had nonfunctional mitochondria, and 1.2% had both a damaged membrane and nonfunctional mitochondria; whereas for frozen-thawed sperm, the percentages of sperm localized in four regions were 74.8%(Region 3), 12.7%(Region 4), 9.9% (Region 2), and 2.6%(Region 1), respectively.

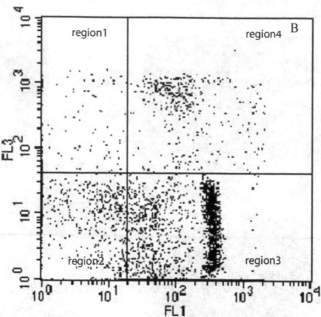

Fig. 4. Flow cytometric dot plots of spermatozoa of red seabream after cryopreservation. Region 1, sperm with a damaged plasma membrane but normal mitochondrial function. Region 2, sperm with an intact plasma membrane but lacking mitochondrial function. Region 3, sperm with an intact plasma membrane and functional mitochondria. Region 4, sperm with a damaged plasma membrane and functional mitochondria.

2.3 Discussion

Motility is an important characteristic for estimating the quality of fresh as well as cryopresrved sperm (Lahnsteiner et al., 1996a). In this study, the freezing-thawing process did not significantly change the main motility pattern and swimming velocity of motile sperm 10 s after activation, and the progressive linear motion was still the dominant pattern. Moreover, for the sperm cryopreserved with 12–21% DMSO, the freezing-thawing process also didn't significantly influence their motility and motility pattern, although it significantly reduced their motility period. However, different results were obtained from the sperm cryopreservation of turbot (Dréanno et al., 1997), which the percentage of motile post-thaw sperm was significantly lower than that obtained from fresh sperm while the velocity and the duration of motion were not significantly modified.

No significant difference in the fertilization rates and hatching rates were observed between sperm cryopreserved with 12–21% DMSO and fresh sperm. However, Lahnsteiner et al. (2003) reported that in cyprinids sperm, the post-thaw fertilization ratios obtained with sperm to egg ratios of $1.3–2.6 \times 10^6{:}1$ did not reach that of the fresh sperm. Similar results have also been reported in turbot (Chen et al., 2004; Suquet et al., 1998) and flounder (Zhang et al., 2003). These may be due to the species specific or un-ideal protocols used in sperm cryopreservation. In this study, for the post-thaw sperm a high positive correlation was observed between the percentage of motile sperm and fertilizing capacity. This was consistent to the results that obtained from turbot (Dréanno et al., 1999), common carp (Linhart et al., 2000) and African Catfish (Rurangwa et al., 2001).

In the present study, the data from ultrastructural investigation and flow cytometric analysis demonstrates that more that 60% of post- sperm were normal in morphology and mitochondrial function. These results further confirmed the high performance of the protocols established for red seabream. In addition, the high fertilization capacity of post-thaw sperm implies that some of the slightly damaged spermatozoa can still fertilize eggs and develop into larvae. However, it remains to be determined whether the larvae from cryopreserved sperm develop into healthy adults.

During the process of cooling, freezing and thawing, spermatozoa are subjected to a series of damages (Oehninger et al., 2000). In ultrastructural investigation, we found 20.7 ± 3.1% were slightly damaged in some way and 16.4 ± 4.2% were severely damaged. One of the causes may be the ice crystal formation during the freezing process and some researchers agree that intracellular ice formation is the major injury mechanism at rapid cooling rates (Toner et al.,1993; Chao & Liao 2001). Other causes of cryodamages include pH fluctuation, cold shock, osmometric effect, and cryoprotectant toxicity (Chao & Liao 2001). The swelling and rupture of the plasmalemma after thawing may be due to the damage to the unit membrane which is very sensitive to freezing and thawing (Lahnsteiner et al., 1992). Similar morphological changes were reported in post-thaw sperm of ocean pout (Yao et al., 2000), rainbow trout (Lahnsteiner et al., 1996c), and atlantic croaker (Gwo et al., 1991). For example, in grayling sperm (Lahnsteiner et al., 1992), a marked decrease in sperm quality was observed, about 40% to 50% of the spermatozoa were completely damaged, 30% to 40% changed and only 10–20% showed an intact morphology. In this study, flow cytometric analysis, based on membrane integrity and mitochondrial function, was used to assess post-thaw sperm quality. After double staining with Rh123 and PI, we found 74.8% of post-thaw sperm showed membrane integrity and mitochondrial function. In rainbow trout (Ogier de Baulny et al., 1997), the plasma membrane and mitochondrial function were better protected with 10% DMSO.

Plasma membrane integrity and mitochondrial function are the two most important attributes for fertilizing an egg. The damage to membrane integrity and mitochondrial function could destabilize the sperm membrane and affect mitochondrial energy metabolism, thereby affecting spermatozoa viability. However, what interested us most is that although about 30% of spermatozoa were damaged in some way or even totally ruptured, the fertilization capacity of post-thaw sperm were not affected significantly in the standardized artificial fertilization experiment (Oehninger et al., 2000). Three hypotheses can be considered. The first hypotheses is that the sperm that survived freezing-thawing with normal morphology and mitochondrial function as shown in Fig. 3 and Fig. 4 region 3 should be similar to fresh sperm in fertilization capacity. The second is that the sperm cryopreserved with the established method could provide adequate numbers of motile spermatozoa with normal sperm parameters to fertilize the eggs in artificial fertilization experiment. The third is that the process of freezing-thawing may result in a population of partially damaged yet motile spermatozoa, which can fertilize eggs and develop into larva normally. Such a population usually exhibits a certain degree of plasma and mitochondrial membrane leakiness as shown in Fig. 3 and Fig. 4 region 2, 4.

3. Conclusion

In conclusion, the fertilizing capacity and egg hatchability were not significantly reduced by the post-thaw sperm treated with 12-21%DMSO, although the post-thaw sperm quality was influenced during the freezing and thaw process in motility, ultrastructure and mitochondrial function. The cryopreservation protocol used for red seabream sperm should be of great value for the establishment of sperm banks and assessment of ultrastructure and flow cytometry facilitated identification of damaged sperm; However, the exact nature of cryodamage to fish sperm are not yet fully understood. Sperm motility, structure integrity and mitochondrion function were damaged with different extent, although the fertilization capacity of cryopreserved sperm was not changed. There are many questions need to answer, how does the cryodamage reduce the sperm motility duration? If the cryodamages influence the gene expression and the embryo and larvae development? how to improve the post-thaw sperm quality by optimize the cryopreservation method?

4. Acknowledgment

This study was funded by the State 863 High-Technology R&D Project of China (2003AA603510 and 2004AA603310), the Knowledge Innovation Program of the Institute of Oceanology, Chinese Academy of Sciences (Y02507101Q), The National Natural Science Foundation of China (NSFC) (41076100 an31072212) and the earmarked fund for Modern Agro-industry Technology Research System (nycytx-50).

5. References

Babiak, I.; Glogowski, J.; Brzuska, E. & Adamek, J. (1997). Cryopreservation of sperm of common carp, Cyprinus carpio L. Aquaculture Research 28(7): 567–571.

Billard, R.; Cosson, J. & Crim, L. W. (1993). Motility of fresh and aged halibut sperm. Aquatic Living Resources 6(1):67–75.

Blaxter, J. H. S. (1953). Sperm storage and cross-fertilization of spring and autumn spawning herring. Nature 172(4391):1189–1190.

Cabrita, E.; Robles, V.; Alvarez, R. & Herráez, M. P. (2001). Cryopreservation of rainbow trout sperm in large volume straws: application to large scale fertilization. Aquaculture 201(3-4):301–314.

Chao, N. H. & Liao, I,C. (2001). Cryopreservation of finfish and shellfish gametes and embryos. Aquaculture 197(1-4):161–189.

Chen, S. L.; Ji, X. S.; Yu, G. C.; Tian, Y. S. & Sha, Z. X. (2004). Cryopreservation of sperm from turbot (Scophthalmus maximus) and application to large-scale fertilization. Aquaculture 236(1-4):547–556.

Chen, S. L.; Ji, X. S.; Yu, G. C.; Tian, Y. S.& Sha, Z. X. (2004). Cryopreservation of sperm from turbot (Scophthalmus maximus) and application to large-scale fertilization. Aquaculture 236(1-4):547–56.

Christensen, J. M. & Tiersch, T. R. (1997). Cryopreservation of channel catfish spermatozoa: effect of cryoprotectant, straw size, and formulation of extender. Theriogenology 47(3):639–645.

Conget, P.; Ferna´ndez, M.; Herrera, G. & Minguell, J. J. (1996). Cryopreservation of rainbow trout (Oncorhynchus mykiss) spermatozoa using programmable freezing. Aquaculture 143(3-4):319–329.

Donoghue, A. M.; Garner, D. L.; Donoghue, D. J. & Johnson, L. A. (1995). Viability assessment of turkey sperm using fluorescent staining and flow cytometry. Poult Sci 74(7):1191–1200.

Dréanno, C.; Suquet, M.; Quemener, L.; Cosson, J.; Fierville, F.; Normant, Y. & Billard, R. (1997). Cryopreservation of turbot (Scophthalmus Maximus) spermatozoa. Theriogenology 48(4):589–603.

Dzuba, B. B.; Kopeika, E. F. (2002) Relationship between the changes in cellular volume of fish spermatozoa and their cryoresistance. Cryo Letters 23 (6):353-60.

Graham, J. K.; Kunze, E. & Hammerstedt, R. H. (1990). Analysis of sperm cell viability, acrosoma integrity and mitoxhondrial function using flow cytometry. Biol Reprod 43(1):55–64.

Gravance, C. G.; Garner, D. L.; Miller, M. G. & Berger, T. (2001). Fluorescent probes and flow cytometry to assess rat sperm integrity and mitochondrial function. Reprod Toxicol 15(1):5–10.

Gwo, J. C. (2000). Cryopreservation of sperm of some marine fishes. In: T. R. Tiersch & P. M. Mazik , (Ed.), Cryopreservation in aquatic species, World Aquaculture Society, Baton Rouge, Louisiana, USA, pp. 138-60.

Gwo, J. C. 2000. Cryopreservation of sperm of some marine fishes. in: T. R. Tiersch & P. M. Mazik, editors. Cryopreservation in aquatic species. World Aquaculture Society, Baton Rouge, Louisiana, USA, pp. 138-160

Gwo, J. C.; Strawn, K.; Longnecker, M. T. & Arnold, C. R. (1991). Cryopreservation of Atlantic croaker spermatozoa. Aquaculture 94(4):355–375.

He, S. & Woods, III. L. C. (2004). Changes in motility, ultrastructure, and fertilization capacity of striped bassMorone saxatilis spermatozoa following cryopreservation. Aquaculture 236(1-4):677–686.

Henderson-Arzapalo, A.; Colura, R. L. & Maciorowski, A. F. (1994). A comparison of black drum, red drum, and their hybrid in saltwater pond culture. Journal of the World Aquaculture Society 25(2):289–296.

Kime, D. E.; Ebrahimi, M.; Nysten, K.; Roelants, I.; Rurangwa, E.; Moore, H. D. M. & Ollevier, F. (1996). Use of computer assisted sperm analysis (CASA) for monitoring the effects of pollution on sperm quality of fish; application to the effects of heavy metals. Aquatic Toxicology 36(1-4): 223–227.

Kopeika, E.; Kopeika, J.; Zhang, T. (2007) Cryopreservation of fish sperm. Methods Mol Biol.;368:203-17.

Kopeika, J., Kopeika, E.; Zhang, T.; Rawson, D. M. (2003a) Studies on the toxicity of dimethyl sulfoxide, ethylene glycol, methanol and glycerol to loach (Misgurnus fossilis) sperm and the effect on subsequent embryo development. Cryo Letters. 24(6):365-74.

Kopeika, J., Kopeika, E.; Zhang, T.; Rawson, D. M.; Holt, W. V. (2003b) Detrimental effects of cryopreservation of loach (Misgurnus fossilis) sperm on subsequent embryo development are reversed by incubating fertilised eggs in caffeine. Cryobiology46(1):43-52.

Kopeika, J.; Kopeika, E.; Zhang, T.; Rawson, D. M.; Holt, W. V. (2004) Effect of DNA repair inhibitor (3-aminobenzamide) on genetic stability of loach (Misgurnus fossilis) embryos derived from cryopreserved sperm. Theriogenology 61(9):1661-73.

Lahnsteiner, F.; Berger, B. & Weismann, T. (2003). Effects of media, fertilization technique, extender, straw volume, and sperm to egg ratio on hatchability of cyprinid embryos, using cryopreserved semen. Theriogenology 60(5):829–841.

Lahnsteiner, F.; Berger, B.; Horvath, A.; Urbanyi, B. & Weismann, T. (2000). Cryopreservation of spermatozoa in cyprindid fishes. Theriogenology 54(9):1477–1498.

Lahnsteiner, F.; Berger, B.; Weismann, T. & Patzner, R.A. (1996a). Physiological and biochemical determination of rainbow trout, Oncorhynchus mykiss semen quality for cryopreservation. J Appl Aquacult;6:47–73.

Lahnsteiner, F.; Berger, B.; Weismann, T. & Patzner, R.A. (1998). Determination of semen quality of the rainbow trout, Oncorhynchus mykiss, by sperm motility, seminal plasma parameters, and spermatozoal metabolism. Aquaculture 163(1-2):163–181.

Lahnsteiner, F.; Berger, B.; Wiesmann, T. & Patzner, R. A. (1996c). Changes in morphology, physiology, metabolism, and fertilization capacity of rainbow trout semen following cryopreservation. Prog Fish Cult 58(3):149–59.

Lahnsteiner, F.; Weismann, T. & Patzner, R. A. (1992). Fine structural changes in spermatozoa of the grayling, Thymallus thymallus (Pisces:-Teleostein), during routine cryopreservation. Aquaculture 103(1):73–84.

Lahnsteiner, F.; Weismann, T. & Patzner, R.A. (1996b). Cryopreservation of semen of the grayling (Thymallus thymallus) and the Danube salmon (Hucho hucho). Aquaculture 144(1-3):265–274.

Linhart, O.; Rodina, M. & Cosson, J. (2000). Cryopreservation of sperm in common carp Cyprinus carpio: sperm motility and hatching success of embryos. Cryobiology 41(3):241–250

Liu, Q. H.; Chen, Y. K.; Xiao, Z. Z.& Li, J. (2010a). Effect of storage time and cryoprotectant concentration on the quality of cryopreserved sperm in red seabream (Pagrus major). Aquaculture research. 41(9):e89-e95.

Liu, Q. H.; Chen, Y. K.; Xiao, Z. Z.; Li, J. & Xu, S. H. (2010b). Effect of long-term cryopreservation on physiological characteristics, antioxidant activities and lipid peroxidation of red seabream (Pagrus major) sperm. Cryobiology 61(2):189-193.

Liu, Q. H.; Li, J.; Xiao, Z. Z., Ding, F. H.; Yu, D. D. & Xu, X. Z. (2007b). Use of computer-assisted sperm analysis (CASA) to evaluate the quality of cryopreserved sperm in red seabream (Pagrus major). Aquaculture 263(1-4):20–25.

Liu, Q. H.; Li, J.; Zhang S. C.; Ding, F. H.; Xu, X. Z.; Xiao, Z. Z. & Xu, S. H. (2006). An Efficient Methodology for Cryopreservation of spermatozoa of Red Seabream, Pagrus major, with 2-mL Cryovials. Journal of the world aquaculture society 37(3):289-297.

Liu, Q. H.; Li, J.; Zhang, S. C.; Xiao, Z. Z.; Ding, F. H.; Yu, D. D. & Xu, X. Z. (2007a). Flow cytometry and ultrastructure of cryopreserved red seabream (Pagrus major) sperm. Theriogenology 67(6):1168–1174.

Oehninger, S.; Duru, N. K.; Srisombut, C. & Morshedi, M. (2000). Assessment of sperm cryodamage and strategies to improve outcome. Mol Cell Endocrinol 169(1-2):3–10.

Ogier de Baulny, B.; Labbe´, C. & Maisse G. (1999). Membrane integrity, mitochondrial function, ATP content, and motility of the European catfish (Silurus glanis) testicular spermatozoa after freezing with different cryoprotectants. Cryobiology 39(2):177–184.

Ogier de Baulny, B.; Le Vern, Y.; Kerboeuf, D. & Maisse, G. (1997). Flow cytometric evaluation of mitochondrial function and membrane integrity in fresh and cryopreserved rainbow trout (Oncorhynchus mykiss) spermatozoa. Cryobiology 34(2):141–149.

Ohta, H.; Kawamura, K.; Unuma, T. & Takegoshi, Y. (2001). Cryopreservation of the sperm of the Japanese bitterling. Journal of Fish Biology 58(3):670–681.

Richardson, G. F.; Wilson, C. E.; Crim, L.W. & Yao, Z. (1999). Cryopreservation of yellowtail flounder (Pleuronectes ferrugineus) semen in large straws. Aquaculture 174(1-2):89–94.

Rurangwa, E.; Volckaert, F. A. M.; Huyskens, G.; Kime, D. E. & Ollevier, F. (2001). Quality control of refrigerated and cryopreserved semen using computer-assisted sperm analysis (CASA), viable staining and standardized fertilisation in African catfish (Clarias gariepinus). Theriogenology 55(3):751–769.

Segovia, M.; Jenkins, J. A.; Paniagua-Chavez, C. & Tiersch, T. R. (2000). Flow cytometric evaluation of antibiotic effects on viability and mitochondrial function of refrigerated spermatozoa of Nile Tilapia. Theriogenology 53(7):1489–1499.

Suquet, M.; Dréanno, C.; Petton, B.; Norman, Y.; Omnes, M. H. & Billard, R. (1998). Long-term effect of the cryopreservaton of turbot (Psetta maxima) spermatozoa. Aquatic Living Resources 11(1):45–48.

Tiersch, T. R.; Figiel, C. R.; Jr.; Wayman, W. R.; Williamson, J. H.; Carmichael, G. J. & Gorman, O. T. (2000). Cryopreservation of sperm of the endangered razorback sucker. In: T. R. Tiersch & P. M. Mazik, (Ed.), Cryopreservation in aquatic species. World Aquaculture Society, Baton Rouge, Louisiana, USA, Pages 117 – 122

Tiersch, T. R.; Wayman, W. R.; Skapura, D. P.; Neidig, C. L. & Grier, H. J. (2004). Transport and cryopreservation of sperm of the common snook, Centropomus undecimalis (Bloch). Aquaculture Research 35(3):278–288.

Toner, M.; Cravalho, E. G. & Karel, M. (1993). Cellular response of mouse oocytes to freezing stress: prediction of intracellular ice formation. J Biomech Eng 115(2):169–174.

Van derWalt, L. D.; Van der Bank, F. H. & G. J. Steyn. (1993). The suitability of using cryopreservation of spermatozoa for the conservation of genetic diversity in African catfish (Clarias Gariepinus). Comparative Biochemistry Physiology 106(2):313–318.

Viveiros, A. T. M.; So, N. & Komen, J. (2000). Sperm cryopreservation of African catfish, Clarias garepinus: cryoprotectants, freezing rates and sperm:egg dilution ratio. Theriogenology 54(9):1395–1408.

Warnecke, D. & Pluta, H. J. (2003). Motility and fertility capacity of frozen/thawed common carp (Cyprinus carpio L.) sperm using dimethyl-acetamide as the main cryoprotectant. Aquaculture 215(1-4):167–85.

Yao, Z.; Crim, L. W.; Richardson, G. F. & Emerson, C. J. (2000). Motility, fertility and ultrastructural changes of ocean pout (Macrozoarces americanus L.) sperm after cryopreservation. Aquaculture 181(3-4):361–375.

Zhang, Y. Z.; Zhang, S. C.; Liu, X. Z.; Xu, Y. Y.; Wang, C. L.; Sawant, M. S.; Li, J. & Chen, S
    L. (2003). Cryopreservation of flounder (Paralichthys olivaceus) sperm with a
    practical methodology. Theriogenology 60(5):989–996.

3

Sperm Cryopreservation of Two European Predator Fish Species, the Pikeperch (*Sander lucioperca*) and the Wels Catfish (*Silurus glanis*)

Zoltán Bokor, Béla Urbányi, László Horváth,
Tamás Müller and Ákos Horváth
Department of Aquaculture, Szent István University
Hungary

1. Introduction

Wels catfish (*Silurus glanis*) and pikeperch (*Sander lucioperca*) are two predator fish species cultivated in the traditional Central European pond aquaculture. Their role in the pond ecosystem is the control of the populations of smaller wild fish that enter ponds during their flooding in the Spring and would represent food competition for the cultured cyprinids. In addition, both species are highly priced for their excellent boneless meat, therefore, attempts are made to improve their culture and enhance yields (Horváth et al., 2002).

The process of induced propagation of wels catfish is based on a long-standing technology. However, certain problems can still appear: the method of collecting male gametes is still based on the removal of testes. By the application of this method a particular male can only be used once for propagation. Moreover, differentiation of sexes requires a great deal of experience and the danger of using an immature female with a less developed body structure is still present. As a result of listed problems the success of propagation becomes questionable.

In recent years successful experiments have been carried out in the pikeperch in many farms and research centers for the development of induced propagation in hatcheries. Synchronization of maturation of female individuals is not perfectly developed yet, thus, successful striping requires a constant attention and control. That is why minimizing the presence of males and securing sperm in a most simple way could focus attention on females.

Application of cryopreserved male gametes for fish propagation in hatcheries can serve as a solution for all mentioned difficulties and risks. History of fish sperm cryopreservation dates back to the beginning of 1950s, since then the sperm of more than 200 fish species has been cryopreserved successfully all over the world (Rana, 1995). In spite of this the application of cryopreserved fish sperm is still not very common in aquaculture in contract to e.g. the dairy cattle sector. Most studies done on cryopreservation of fish sperm put the emphasis on optimization of the process, on cryopreservation of a small amounts of male gametes, thus, rendering this technology to the level of laboratory experiments without any basic output for farmers.

By the help of a successful cryopreservation method applicable in fish farms not only the reduction of propagation risks would become possible but also a long-term storage of gametes of substantial breeders as already applied in case of carp. Development of a sperm bank already employed at cattle breeding would also become feasible thus increasing the role and rate of rarely used selection methods of animal breeding in fisheries.

The objective of our work was the development of working protocols for the cryopreservation of wels catfish and pikeperch sperm that can be applied to aquaculture of these species. In both species, this development required a thorough knowledge of culture conditions of the given species, studies of the cryoresistance of sperm and adaptation of cryopreservation methods to hatchery practices.

2. Materials and methods

Experiments on the cryopreservation of sperm of both species were carried out during the years 2005-2008 at various locations in Hungary. Details of experiments including their dates and locations are provided in the descriptions of experiments on each species.

2.1 Experiments on wels catfish sperm

2.1.1 Methods of collection and cryopreservation of male gametes

Gametes applied in the experiments were invariably retrieved directly from the testis removed from the abdomen after the decapitation of male wels catfish (not stripped). After removing it from the abdomen the testis was cut up and squeezed out through a dry gauze into a Petri-dish. After extraction of sperm motility of gametes was examined through a light microscope at 200× magnification.

Ten per cent methanol was used as a cryoprotectant and 6% fructose as a diluent. pH of the cryopreservation medium was adjusted to 7.73 by the help of 1 M $NaHCO_3$ solution. From the diluted sperm treated this way 4 ml was pipetted into a 5-ml straw.

In the process of cryopreservation liquid nitrogen was poured into a polystyrene box on the top of which a polystyrene frame was placed with a height of 3 cm and the straws were laid on it. Samples were stored in a canister storage dewar until use.

Straws were thawed in 40 °C of water for 40 seconds. After thawing the closed ends of straws were cut up and their content was poured into a test-tube or directly onto the eggs used for propagation. Motility of thawed sperm was examined according to the method already described. The method mentioned here was compiled on the basis of former experiments of the cryopreservation group of the Department of Aquaculture on African catfish (Urbányi et al., 1999; Horváth & Urbányi, 2000).

2.1.2 Experiments on cooling time and sperm-egg ratio

Male gametes applied for cryopreservation in 2005 were collected from the farm in Tuka of Szarvas-Fish Kft. and the farm in Szeged of Szegedfish Kft. by joining to their propagation processes. Males were injected with 4 mg/body weight kg carp pituitary in one dose by the assistants of the farm before the extraction of sperm. Length and weight of the body and weight of the testis were measured. Gonadosomatic indices (GSI) were determined from the ratio of testis and body weight. After the squeezing of sperm the motility of gametes was defined as described above. After the collection of gametes the former described cryopreservation method was used with the addition that in case of samples collected in Tuka the effect of cooling time on motility and fertilizing capacity was also tested. Cooling time of samples varied between 3, 5 and 7 minutes. After the cooling period straws were placed into liquid nitrogen.

First propagation tests were completed in the Szajol farm of Fish-Coop Kft. Eggs were gained from fish by a routine propagation process. In the first experiments eggs were divided into 40, 80 and 120 g doses and each dose was propagated with 1 straw of cryopreserved sperm. Fertilized eggs were then incubated in 7 l Zug-jars. Hatching rate was counted after hatching.

The second experimental procedure was performed in the Attala fish farm of Attala Hal Kft., when eggs were divided up to two 150 g doses and one of them was fertilized with one, while the other one with two straws of sperm. In both cases freezing time was 7 minutes. At hatching the ratio of hatching and deformed larvae was determined.

In the experimental procedure the cooling rate was measured, too. A straw was filled with cryopreservation medium. The K type sensor of a Digi-Sense DualLogR digital thermometer (Eutech Instruments, Singapore) was placed into the straw which was then laid onto a 3 cm high polystyrene frame floating on the surface of liquid nitrogen. The thermometer recorded temperature data with 1 second intervals. Temperature data were collected for 6 minutes since storing capacity of the memory of the thermometer allowed the recording of this amount of data.

2.1.3 Cryopreservation and analysis of sperm collected outside of the spawning season in hatchery conditions

In 2006 wels catfish sperm was collected in January and March (aside from the spawning season) in the Köröm farm of Aqua-culture Kft. (Köröm Fish Farm, Local Government of Bőcs at present) from wels catfish kept in a flow-through intensive system. They were kept in tanks in a constant water temperature of 20 °C. The method applied for cryopreservation was the same as already mentioned with the difference that sperm was frozen for 7 minutes on the polystyrene frame before placing it into liquid nitrogen. Motility of sperm was examined both in fresh and cryopreserved samples.

Propagation experiments were performed at the Szajol farm of Fish-Coop Kft. For this, cryopreserved samples originating from Szeged, 2005 and Köröm, 2006 were used. Eggs were collected from female fish by a routine propagation process. In the experimental procedure eggs were distributed into 250 and 350 g dosages and fertilized with one straw of sperm. Fertilized egg doses were then incubated in 7 l Zug-jars and hatching rates were determined after hatching.

2.1.4 Application in hatcheries

The aim of these experiments was to fertilize significant amounts of eggs (150-300 g) with large doses of cryopreserved sperm (5-ml straws) all over the country by joining to propagation work of a given farm. Reliability and repeatability of the method were also examined. In each case one dose of eggs was fertilized with the content of one straw. In the frames of a routine propagation work fertilization with cryopreserved sperm was propagated in five different farms:

- in Attala pond farm of Attala Fish Production and Trading Kft.
- in Köröm fish farm of the Local Government of Bőcs
- in Százhalombatta farm of TEHAG Kft.
- in Ördöngös farm of Aranykárász Co.
- in Szeged farm of Szegedfish Kft.

In the Attala experiment in 15 May 2007 cryopreserved sperm collected ourside of the spawning season was used for the fertilization of 200 g egg doses.

The experiment in Köröm was carried out in 17 May 2007 in which cryopreserved sperm was applied for fertilization also collected off season. Egg doses of 200 g were fertilized both for treated and for control groups.

Following that, egg doses of 200 g were fertilized in the Százhalombatta farm of TEHAG Kft. in 22 May 2007. This time cryopreserved sperm from Szeged was used in the experiments collected in 2006 in the spawning season.

In the fourth experiment in 21 May 2007 egg doses of 200-350 g were fertilized in the Ördöngös farm of Aranykárász Co with cryopreserved sperm collected off season.

The last experiment was performed in the hatchery of Szegedfish Kft. in 23 May 2007. In the research work egg doses of 150 g were fertilized with cryopreserved sperm deriving from Szeged.

In all cases our team personally joined to the propagation work of the farm, thus, we ensured that an adequate amount of cryopreserved sperm was used for the egg doses. Process of propagation was always performed according to the practice of the certain farm. One dose of eggs was fertilized with one straw of thawed sperm. In all experiments hatching percentage of larvae was determined and in the experiment in Attala fertilization rate at 4-8 cell stage was also examined.

2.1.5 Experiments on larval survival

For the analysis of larval survival, growth and survival of feeding larvae in 2007 and non-feeding wels catfish larvae in 2008 were examined.

Analysis of feeding larval stage

When examining feeding larval stage in farm conditions, larvae were produced in Százhalombatta farm of TEHAG Kft. by applying local propagation methods. During fertilization process 1 straw of sperm was added to 200 g of eggs. After fertilization the 200 g doses of eggs were placed into 7 1 Zug-jars. For fertilization of individuals in the control group native sperm from males of the farm was used. On the second day after fertilization hatched larvae were counted then after hatching the non-feeding larvae were placed into larva-tanks. On the third day after hatching when larvae started their exogenous feeding the ones devoted for the experiment were counted and placed into troughs. 1000 individuals were placed into a 100 1 trough with flow-through water in 3 replcates. The stock was fed

Sperm Cryopreservation of Two European Predator Fish Species, the Pikeperch (Sander lucioperca) and the Wels Catfish (Silurus glanis)

45

with chopped tubifex in every 3 hours. To prevent infections 36% formalin treatment was applied in a concentration of 10 ml/trough in every 4 hours. In each trough velocity of water flow-through was 3 l/minute. Rearing in the trough lasted for 10 days according to the routine practice of the farm.

Laboratory experiments were performed in the Department of Aquaculture. Troughs of the recirculation system applied in the experiment were 40 cm long, 15 cm deep and 10 cm wide (though water depth was 10 cm due to the outlet/stub). Due to photophobia of wels catfish larvae the system was located in a dark room. In the experiment 5×100 individuals (larvae hatched after the application of control and cryopreserved sperm) took part in the treatments. In the first 4 days fish were fed once in every three hours. At the morning and evening feeding they were fed with plankton while at other feeding times experimental fish were fed with Perla Proactiv 4.0 fish diet. Following this 4-day period fish were fed with the above mentioned diet (sometimes also with plankton or Artemia) 4 times a day. Velocity of water flow-through in the troughs was 0.25 l/minute.

In addition to body length and weight condition factor, specific growth rate (S.G.R.) and survival rate (%) was measured.

Examination of non-feeding larval stage

In this developmental stage no farm study was done due to the fact that the trough system was not suitable for the accommodation of such small larvae. Moreover, the hatchery protocol for fry rearing does not use the method of rearing in troughs at this age.

Larvae examined at non-feeding life stage were produced in June 2008 in the hatchery of TEHAG Kft. The method of propagation and utilization of cryopreserved sperm was the same as described at the examination of feeding larval stage. After hatching larvae were transported to the laboratory of the Department of Aquaculture in Gödöllő where the survival of fish was tested using the rearing system built in the previous year. The experiment lasted for 4 days.

2.1.6 Experiments on larval survival

Main effects of cooling time and the quantity of eggs on hatching rate was examined in the Szajol experiment (Part 2.1.2) by the help of two-way analysis of variance ($P < 0.05$). In case of testis collected off season at two sampling times as described in Chapter 2.1.3 GSI values were compared by the help of a two-sample t-test. Hatching results gained from fertilization experiments with cryopreserved and control sperm described in Chapter 2.1.4 were analysed with a t-test, too. These statistical tests were performed using the software GraphPad Prism 4.0 for Windows. In the examinations of larval rearing results of survival were compared with Chi²-test (Kruskal Wallis test), while body length, body weight, condition factors and SGR values with t-tests (Chapter 2.1.5) using the software SPSS 13 for Windows.

2.2 Experiments on pikeperch sperm

2.2.1 Origin of fish

Experiments on pikeperch were conducted two times and in two places: the first place was the Keszthely-Tanyakereszt fish laboratory of Georgikon Faculty of Agriculture at Pannon University and the second one was the Attala hatchery of Attala Hal Kft.

2.2.2 First experiment

The stock of pikeperch taking part in the first experiment originated from Aranyponty Kft. (Sáregres-Rétimajor) (8 females and 10 males: 1424-1870 g). Males were anesthetized by clove oil then sperm was manually stripped and collected with an automatic pipette. Care was taken not to contaminate the gametes with urine or feces.

Motility of fresh sperm was estimated after activating it with water. Motility was examined on slides in a 20× dilution at 200× magnification by the help of a Zeiss Laboval 4 microscope (Carl Zeiss, Jena, GDR). Density of sperm was examined by the Bürker-chamber method in a 1000× dilution.

The following diluents were prepared:

- Glucose diluent (350 mM glucose, 30 mM Tris, pH 8,0)
- KCl diluent (200 mM KCl, 30 mM Tris, pH 8,0)
- Sacharose diluent (300 mM sacharose, 30 mM Tris, pH 8,0)

Methanol and dimethyl-sulfoxide (DMSO) were used as cryoprotectants in a concentration of 10%. All chemicals used in the experiments were purchased from Reanal Zrt. (Budapest, Hungary).

Sperm (200 µl) mixed with a cryopreservation medium (200 µl cryoprotectant, 1600 µl diluent) in a ratio of 1:9 (Horváth et al., 2003; Urbányi et al., 2006) was pipetted into individually marked 0.5-ml straws after 3 minutes of equilibration time. Samples were cryopreserved in a polystyrene box filled with liquid nitrogen. A 3 cm high polystyrene frame was placed onto the top of the nitrogen then straws were laid on this frame where temperature was around -165°C. The time of cryopreservation was 3 minutes. After the freezing process straws were placed into liquid nitrogen and stored there until being used. Thawing was carried out in a 40°C water bath for 13 seconds (Horváth et al., 2003; Horváth et al., 2005). After thawing sperm motility was examined with the same method as described at fresh sperm.

Eggs were distributed to Petri-dishes with a diameter of 5 cm with 200-350 eggs/dose. Fertilization was made with thawed sperm of a half straw (250 µl) right after taking the straw out of the water. Sperm was poured on the egg doses then the gametes were activated with 1 ml of water. Next eggs were allowed to stick to the bottom of the Petri-dish by taking care of the eggs being located in one layer. Fertilization rate was counted at neurula stage.

2.2.3 Second experiment

The second experiment was performed in line with pikeperch propagation in a hatchery. In this research sperm of 4 males and eggs of 1 female were applied. In this case only glucose was used as a diluent with 10% concentration of methanol and DMSO cryoprotectants. Sperm was diluted in a 1:1 and 1:9 ratios. The process of cryopreservation and thawing was the same as the process applied in the first experiment.

Eggs were divided into 10 g doses (about 10 000 eggs according to my counts) and taken into plastic bowls. Each dose was fertilized by a thawed straw of samples (0.5 ml) then these doses were placed into 7 l Zug-jars until hatching. Finally hatched larvae were counted and hatching rate was defined.

Sperm Cryopreservation of Two European Predator Fish Species, the Pikeperch (Sander lucioperca) and the Wels Catfish (Silurus glanis)

47

2.2.4 Application of cryopreserved sperm to hatchery conditions (preliminary experiment)

This research was done in April 2007 in the hatchery of Attala Hal Kft. Gametes used in the experiment originated from the same farm. Male and female fish were treated by the method applied in the farm in line with local hatchery propagation. In the previous experiment it was difficult to avoid mixing of sperm with urine so in this case stripping was performed by a silicon catheter (inside diameter: 1 mm, outside diameter: 1.5 mm) which was introduced into the sperm duct. Motility estimations were made by the method described earlier. Sperm concentration was examined in a Bürker-chamber in a 1000× dilution.

Sperm originating from 3 males were used in the experiment. It was diluted in a 1:1 ratio with the following composition of cryopreservation medium: 350 mM glucose, 30 mM Tris, pH 8.0 (titrated with ccHCl), methanol with a concentration of 10%. Diluted gametes were pipetted into 0.5-ml straws. Cryopreservation and thawing methods arranged to the method described in the previous chapter. Sperm was stored for 1 week in liquid nitrogen in a canister. After thawing the motility of samples was also examined.

Stripped eggs were divided into 10 and 30 g doses in 3 replicates and into a 50 g dose in one replicate. One dose of eggs was fertilized with one straw. Fresh sperm was used as a control. Each dose was poured into a 7 l Zug-jar for incubation. Hatching rate was counted after hatching.

2.2.5 Applied statistical methods

Results of experiments were evaluated with Graphpad Prism 4.0 for Windows program. Effect of cryoprotectants and diluents on motility and fertilization and the effect of diluent ratio and cryoprotectants on hatching ratio was examined by a two-way analysis of variance (ANOVA) (Chapter 2.2.2).

Results gained from the second experiment (Chapter 2.2.3), namely motility (cryopreserved and fresh sperm) and hatching results (in case of 10 and 30 g egg doses) were analysed by the help of a two-sample t-test ($P \leq 0.05$).

3. Results

3.1 Experiments on wels catfish sperm

The GSI (gonadosomatic index) of male individuals from Tuka was $2 \pm 4\%$. The motility of sperm before cryopreservation was 90%, while after thawing this rate was 0% in case of 3 minutes, 40% in case of 5 minutes and 70% in case of 7 minutes long freezing time. The motility of sperm from Szeged was 80%.

In the experiments carried out in Szajol the highest hatching rate ($51 \pm 1\%$) was observed at 7 minutes freezing time and 40 g of dose of eggs, although in the case of 5 and 7 minutes of freezing time a very similar hatching rate (between $40 \pm 0\%$ and $51 \pm 1\%$) was observed (Figure 1.). Only the cooling time had a significant main effect ($P < 0.0001$) on the results, considering that 3 minutes long cooling time gave lower hatching rate.

Hatching rate of propagated eggs was 94% in the case of fertilization with a single straw in Attala, while fertilization with two straws resulted 77% hatching rate. The control results were 89% and 81%. It is worth to mention that the ratio of deformed larvae hatched from eggs fertilized with a single straw was only 2.4% (1.8% in control), while in the case of fertilization with two straws it was 11.2% (7.3% in control).

The cooling rate of a straw was approximately -23°C/minute (Figure 2.). It was observed that the temperature of the straw was only -45°C after 3 minutes while after 5 minutes it was -104°C.

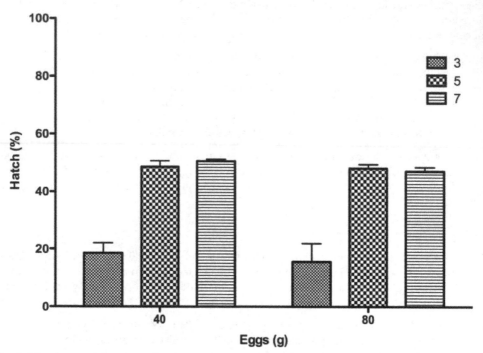

Fig. 1. Hatch rates of wels catfish eggs fertilized with cryopreserved sperm. Each batch of eggs (40 or 80 g) was fertilized with one 5-ml straw of cryopreserved sperm. Columns indicate cooling times employed during cryopreservation (3, 5 or 7 minutes). Data are presented as Mean ± SD (N = 3).

When sperm was collected outside of the spawning season, the average weight of the testes of wels catfish catfish from Köröm was 20.4 g and the average weight of the fish was 2.52 kg, thus GSI rate of them was lower than 1% except one male. This low GSI rate have not had adverse effects on the quality of sperm. No significant difference was observed between GSI rate in January and in March (P = 0.4589). The motility of fresh sperm varied between 50% and 90%. Two of the sperm samples selected for cryopreservation were excluded from further examinations because the motility of these samples was the lowest (50 - 60%). This low rate was caused likely by injuring the cells during squeezing of the testes. Some of the cryopreserved samples were thawed 5 days after freezing and their motility was about 50%.

Sperm frozen in 2005, in Szeged and 2006, in Köröm were used for the propagation experiments. Hatching rate varied between 70 - 80%, except for one sample with 20% of hatching rate. However, according to the head of the farm the fertilization of control group was as bad as the result of the 20% hatching rate. On the basis of the results it was observed that sperm form Köröm (out of spawning season) had similar hatching rate to the sperm from Szeged.

Results of different hatching experiments depended on the propagation method and on the quality of sperm. Fertilization with cryopreserved sperm form Attala resulted 97 ± 1% of fertilization rate while control fertilization rate was 93 ± 1%. There was no significant difference between the two rates (P = 0.0084). In the same experiment the hatching rate of the larvae was 95 ± 2% while in the control group this rate was 94 ± 6%. There was no statistically significant difference between the hatching rate of larvae originating from cryopreserved or fresh sperm.

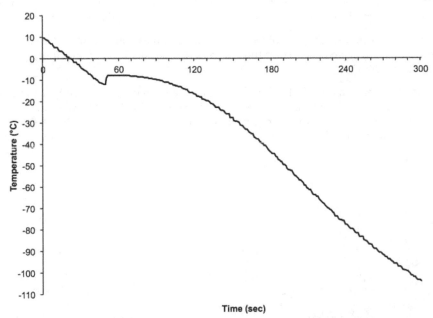

Fig. 2. The cooling profile used with 5-ml straws in the experiments on wels catfish sperm

The hatching rate of larvae originating from cryopreserved sperm in Köröm was 84 ± 5%, while this rate in case of larvae originating from fresh sperm was 69 ± 16%. There was no significant difference between the results. It was observed in the experiments carried out in Százhalombatta that hatching rate of larvae originating from cryopreserved sperm was 50 ± 3%, while the result of the control group was 50 ± 6%. There was also no statistically significant difference between the two groups. Hatching results of the experiments carried out in Ördöngös at the place of Aranykárász Kft. were about 57 ± 22% in case of larvae originating from cryopreserved sperm and 22 ± 18% in case of larvae originating from fresh sperm. In this experiment a significant difference was found (P = 0.05) in favor of the cryopreserved group. The hatching rate of larvae originating from cryopreserved sperm in Szeged was 75 ± 3%, while this rate in case of the control group was 83 ± 1%. These results also differ significantly (P = 0.0249) but now in favor of the control group.

In the experiments on larval survival, statistically significant difference (P = 0.034) was observed on feeding larvae in laboratory conditions regarding 10-day body length. The results showed that larvae originating from cryopreserved sperm had a longer body. During non feeding larval period final body length (P < 0.001) and final weight (P = 0.018) differed significantly in the two groups in favor of larvae originating from cryopreserved sperm. There was no difference between the larvae originating from cryopreserved or fresh sperm in terms of larval survival.

3.2 Experiments on pikeperch sperm

In the first experiment, in spite of all efforts sperm was mixed with urine during stripping, thus the motility of pikeperch sperm was 50 ± 17%. The motility of the best thawed sample was 28 ± 21 %, which was cryopreserved with glucose diluent and DMSO as cryoprotectant, but statistically significant difference was not be observed among the treatmenrts.

The density (spermatozoa/ml) of pikeperch sperm samples counted in a Burker chamber were the following: 1st male: 0.9375×10^{10}, 2nd male: 1.0100×10^{10}, 3rd male: 0.7037×10^{10}, 4th male: 0.6687×10^{10}.

The highest fertilization rate (43 ± 12%) was observed also in the case of using a combination of glucose diluent and DMSO as cryoprotectant (Figure 3.). During statistical analysis of the data it was found that only the cryoprotectant had a significant effect (P = 0.0338) on the ratio of the fertilization.

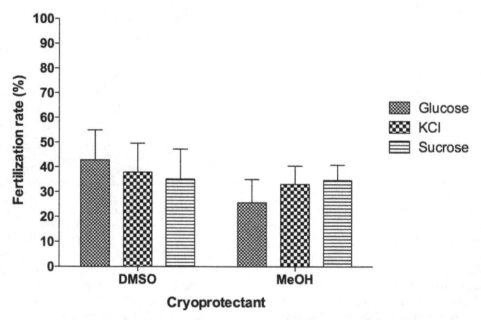

Fig. 3. Fertilization percentages of pikeperch eggs fertilized with cryopreserved sperm in the first experiment. The cryoprotectants dimethyl-sulfoxide (DMSO) and methanol (MeOH) and glucose, KCl or sucrose extenders were compared. Data are expressed as Mean ± SD (N = 3).

The volume of the sperm stripped from pikeperch males in the second experiment was very
ow (less than 1 ml/individual). The motility of fresh sperm was $45 \pm 30\%$. Similarly to the
previous experiment the sperm was mixed with urine again. Motility of thawed pikeperch
sperm was very low (0 – 2%) in the samples containin the cryoprotectant DMSO, while
motility of sperm frozen in presence of methanol was 40%, independently from rate of
dilution. The highest hatching rate ($41 \pm 22\%$) was observed with the use of methanol and
1:1 dilution rate, although the statistical analysis has not shown significant differences
between hatching rates (Figure 4.).

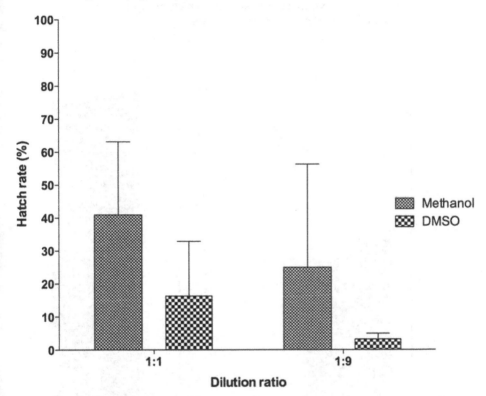

Fig. 4. Fertilization rates of pikeperch eggs fertilized with cryopreserved sperm in the
second experiment. The cryoprotectants methanol and dimethyl-sulfoxide (DMSO) and
dilution ratios of 1:1 and 1:9 were compared. Data are expressed as Mean $\pm$ SD (N = 3).

In the hatchery experiment Stripping sperm with silicon catheter resulted that the motility
was $63 \pm 10\%$. Concentration of sperm was $1.8571 \pm 0.1538 \times 10^{10}$, while the number of
eggs/g was 1367 ± 54, thus the number of sperm for an egg was 3.396×10^5 in the case of 10-
g dose of eggs, 1.132×10^5 in the case of 30-g dose of eggs and 6.792×10^4 in the case of 50-g
dose of eggs. Motility of sperm after thawing was $53 \pm 5\%$, thus there was no significant
difference (P = 0.1135) between the motility of fresh and cryopreserved sperm.

Fertilization of the dose of 10 g of eggs with a single straw resulted $47 \pm 4\%$, while in the
case of the dose of 30 g eggs resulted $55 \pm 3\%$ hatching rate (Figure 5.). There was no

statistically significant difference between the results of the different doses however the result of t-test (P = 0.05701) was very close to the significance level. A hatching rate of 87% was observed in the case of fertilization the dose of 50 g eggs with one thawed straw although in this case there were no replicates in the experiment. It was observed, however, that egg batches of different weight behaved differently in the hatching jars. While egg batches of 10 g stuck together in spite of the attempted elimination of egg stickiness, those in batches of 30 g or 50 g freely rolled on each other, thus improving oxygen supply of fertilized eggs and developing embryos. Thus, it is recommended to use larger batches of eggs for fertilization with cryopreserved sperm, which in turn would facilitate the acceptance of this technology in the aquaculture practice.

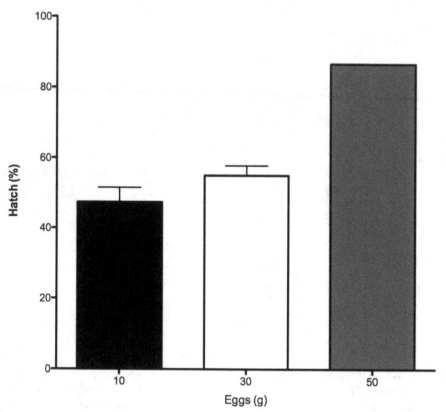

Fig. 5. Hatch rates of pikeperch eggs fertilized with cryopreserved sperm in a commerical hatchery during routine spawning work. Egg batches of 10, 30 or 50 g were used for fertilization (N = 3).

4. Discussion

4.1 Experiments on wels catfish sperm

Hatchery propagation of catfish species including the wels catfish faces several problems. Males of catfish species are typically oligospermic and sperm cannot be stripped but has to be extracted from surgically removed testes (Legendre et al., 1996). As it was mentioned in the Introduction, this can lead to several problems such as shortage of sperm during induced spawning to unnecessary killing of immature females due to their erroneous identification as males.

Cryopreservation of the sperm of catfish species has been studied extensively. Several studies have been published on the cryopreservation of wels catfish sperm (Krasznai & Márián, 1985; Linhart et al., 1993; Ogier de Baulny et al., 1999; Linhart et al., 2005), however, they all reported the use of minute amounts of gametes and did not test practical utilization in the hatcheries.

It can be concluded according to the measured freezing parameters that a longer cooling time is needed for the safe cryopreservation of sperm in 5-ml straws because the temperature is not low enough (-45°C) after 3 minutes. The 7 minutes cooling time, which was used during the experiments, is suitable for these 5-ml straws.

Large amounts of eggs can be fertilised safely with a single 5-ml straw. It was observed in these experiments that the amount of the eggs, fertilised with one straw can be increased because the 2 ml sperm that can be found in a straw contained enough spermatozoa to fertilise 120 g eggs. The use of 5-ml straws has been tested on several fish species including the rainbow trout *Oncorhynchus mykiss* (Wheeler & Thorgaard, 1991; Lahnsteiner et al., 1997; Cabrita et al., 2001) or the paddlefish *Polyodon spathula* (Horváth et al., 2010), however, all previous works report a more or less reduced fertilizing capacity of sperm cryopreserved in these straws as compared to the conventional 0.5-ml French straws. The reaction of sperm to cryopreservation in different straw types seems to be species specific with the sperm of the wels catfish being especially resistant to the incurred cryodamage.

In the experiments no significant decrease was experienced in the quality of sperm after thawing. One of the reasons of this is that proper cooling time was successfully defined, which resulted the best fertilization rate. Thus, it can be said that cryopreserved sperm does not decrease the hatching rate compared to the traditional, routine method.

According to these experiments wels catfish catfish sperm collected outside of spawning season is as suitable for cryopreservation and for fertilization at fish farms similarly to the ones collected in the spawning time.

After the successful cryopreservation and thawing of large amounts of sperm the next step is to carry out safe fertilization with this sperm on large scale. In the experiments doses of eggs between 150 - 350 g were fertilised with a single straw. According to literature data 100 - 200 g eggs can safely be incubated in one 7-l Zug-jar (Szabó, 2000). Cryopreserved sperm showed similar hatching rates to control in every experiment. These results prove that maximum sperm-egg ratio was not reached that might cause a decrease in hatching rate. According to the experiments it can be said that the improved method is suitable for wels catfish fertilization.

After improving the freezing method of sperm the next task was to examine whether the growth and survival of larvae originated from cryopreserved sperm reaches that of larvae originated from fresh sperm. The research was extended to both the feeding and non-feeding larval periods. The results in both cases were that there is no difference in the survival of larvae fertilised with cryopreserved or fresh sperm. In the non-feeding larval period the growth of larvae from cryopreserved sperm exceeded the growth of the control, and in feeding larvae body length was higher compared to the control results.

According to these experiments the survival rate of larvae originating from cryopreserved sperm is as high as in the control and growth level of them in some cases showed better results compared to the control.

4.2 Experiments on pikeperch sperm

During the improvement of the cryopreservation technique of pikeperch sperm in laboratory cryoprotectant DMSO showed better fertilization rates than methanol but fertilization experiments in hatcheries showed opposite results. Literature data can be found on successful usage of both cryoprotectants in several fish species. The objective of this thesis is the usage of cryopreserved pikeperch sperm in hatcheries and according to the results of the experiments in the whole it was concluded that methanol and 1:1 dilution rate is suitable for freezing pikeperch sperm.

A significant variation was observed in motility after thawing and in hatching rate in the first experiments. This variability is caused by mixing of sperm with urine. This problem can successfully be eliminated when the stripping of sperm is conducted with a silicone catheter. According to this method the sperm is stripped with this silicone catheter directly from the testes preventing the mixing of sperm with urine or feces. One year later the use of catheter resulted in substantially better hatching rates.

It was observed that the increasing of the amount of eggs fertilised with a single 0..5-ml straw resulted in improved hatching rates. The reason for this can be that different amounts of eggs behaved differently in Zug-jars. The dose of 10 g of eggs were slightly stuck together, the dose of 30 g of eggs stuck in smaller batches while the dose of 50 g of eggs rolled freely. In spite of the fact that the 50 g of eggs sample had not replicates, these results suggest that fertilisation of larger amounts of eggs result in better hatching rates.

It is supposed that the eggs in the middle of the 10-g batches were more sensitive for oxygen deficiency than the more loose larger egg samples.

Another explanation for these results is that methanol in smaller eggs samples was in higher concentration, thus the toxic effects were more drastic than in larger samples. The lower sperm-egg ratio in larger egg samples had no influence on the results, suggesting that the amount of sperm was in surplus in the case of smaller egg samples.

5. Conclusion

A method has been developed for the cryopreservation of wels catfish sperm that can be used in the practice of fish farms. It is possible to fertilize 150-300 g eggs with sperm cryopreserved in large volumes (5-ml straws). The motility and hatching rate of frozen sperm correspond with the currently used routine method of fertilization with fresh sperm.

The survival and growth parameters of wels catfish larvae originating from cryopreserved sperm was investigated for the first time. According to this study it can be said that survival rate of larvae originating from cryopreserved sperm reaches and in some cases exceeds that of control larvae. This result proves the practical usage of the cryopreservation method.

Pikeperch sperm has successfully been cryopreserved for the first time, and the developed method was tested in hatchery conditions.

6. Acknowledgment

This work was supported by the NKFP 4/006/2004, RET 12/05 and the Bolyai János Fellowship of the Hungarian Academy of Sciences. The authors express their gratitude to the fish farms mentioned in the work for their support.

7. References

Cabrita, E.; Robles, V.; Alvarez, R. & Herráez, M.P., (2001). Cryopreservation of rainbow trout sperm in large volume straws: application to large scale fertilization. *Aquaculture*, Vol.201, pp. 301–314, ISSN: 0044-8486

Horváth, Á. & Urbányi, B. (2000). The effect of cryoprotectants on the motility and fertilizing capacity of cryopreserved African catfish *Clarias gariepinus* (Burchell 1822) sperm. *Aquaculture Research*, Vol.31, pp. 317-324, ISSN: 1355-557X

Horváth, Á.; Miskolczi, E. & Urbányi, B. (2003). Cryopreservation of common carp sperm. *Aquatic Living Resources*, Vol.16, pp. 457-460, ISSN: 0990-7440

Horváth, Á.; Wayman, W.R.; Urbányi, B.; Ware, K.M.; Dean, J.C. & Tiersch, T.R. (2005). The relationship of cryoprotectants methanol and dimethyl sulfoxide and hyperosmotic extenders on sperm Cryopreservation of two North-American sturgeon species. *Aquaculture*, Vol.247, pp. 243-251, ISSN: 0044-8486

Horváth, L.; Tamás, G. & Seagrave, C. (2002). *Carp and pond fish culture, Second edition*, Blackwell, ISBN 0-85238-282-0, Oxford, UK

Krasznai, Z. & Márián, T. (1985). Kísérletek a lesőharcsa (*Silurus glanis* L.) spermájának mélyhűtéses tartósítására és az évszaktól független szaporításra. *Halászat*, Vol.31, pp. 25-28, ISSN: 0133-1922

Lahnsteiner, F.; Weismann, T. & Patzner, R.A., (1997). Methanol as cryoprotectant and the suitability of 1,2 ml and 5 ml straws for cryopreservation of semen from salmonid fishes. *Aquaculture Research*, Vol.28, pp. 471-479, ISSN: 1355-557X

Legendre, M.; Linhart, O. & Billard, R. (1996). Spawning and management of gametes, fertilized eggs and embryos in Siluroidei, *Aquatic Living Resources*, Vol.9, pp. 59-80, ISSN: 0990-7440

Linhart, O.; Billard, R. & Proteau, J.P. (1993). Cryopreservation of European catfish (*Silurus glanis* L.) spermatozoa. *Aquaculture*, Vol.115, pp. 347-359, ISSN: 0044-8486

Linhart, O.; Rodina, M.; Flajshans, M.; Gela, D. & Kocour, M. (2005). Cryopreservation of European catfish *Silurus glanis* sperm: Sperm motility, viability, and hatching succes of embryos. *Cryobiology*, Vol.51, pp. 250-261, ISSN: 0011-2240

Ogier de Baulny, B.; Labbé, C. & Maisse, G. (1999). Membrane integrity, mitochondrial activity, atp content, and motility of the European catfish (*Silurus glanis*) testicular spermatozoa after freezing with different cryoprotectants, Cryobiology, Vol.39, pp. 177-184, ISSN: 0011-2240

Rana, K.L. (1995). Preservation of Gametes. In: *Broodstock Management and Egg and Larval Quality*, N.R. Bromage & R.J. Roberts, (Eds.), 53-75, Blackwell, ISBN 978-0-632-03591-5, Stirling, UK

Szabó, T. (2000). A csuka tógazdasági tenyésztése In: *Halbiológia és haltenyésztés*, L. Horváth, (Ed.), 312 Mezőgazda Kiadó, ISBN 963-9239-45-3, Budapest, Hungary (in Hungarian)

Urbanyi, B.; Horváth, Á. & Dinnyés, A. (1999). Cryopreservation of African catfish (*Clarias gariepinus*) sperm using different cryoprotectants. *Theriogenology*, Vol.51, p. 296, ISSN: 0093-691X

Urbányi, B.; Szabó, T.; Miskolczi, E.; Mihálffy, Sz.; Vranovics, K. & Horváth, Á. (2006). Successful fertilization and hatching of four European cyprinid species using cryopreserved sperm. *Journal of Applied Ichthyology*, Vol.22, pp. 201-204, ISSN: 0175-8659

Wheeler P.A. & Thorgaard G.H., 1991: Cryopreservation of rainbow trout semen in large straws. *Aquaculture*, Vol.93, pp. 95–100, ISSN: 0044-8486

4

Cryopreservation of the Sperm of the African Catfish for the Thriving Aquaculture Industry in Nigeria

Ofelia Galman Omitogun[1,*], Olanrewaju Ilori[1], Olawale Olaniyan[1],
Praise Amupitan[1], Tijesunimi Oresanya[2],
Sunday Aladele[2] and Wasiu Odofin[2]
*[1]Biotechnology Laboratory, Department of Animal Sciences,
Obafemi Awolowo University (OAU), Ile-Ife,
[2]National Centre for Genetic Resources and Biotechnology (NACGRAB),
Moor Plantation, Ibadan,
Nigeria*

1. Introduction

The production of fish in Nigeria is still very small and cannot sufficiently satisfy the increasing demand of its population of 140 million. To solve the populace's high demand for fish, Nigerians resort to aquaculture which is currently faced with major constraints including lack of fish seed and quality of feed. The scarcity of good broodstock has necessitated the need to conserve the fish genetic resources which are wasted during natural and artificial induced spawning process of fish breeding. One way of expanding aquaculture in Nigeria is by devising a means of preserving genetic resources of our broodstock for all year round supply of fish seed through cryopreservation (Omitogun *et al.*, 2006).

The African catfish *Clarias gariepinus* Burchell, 1822 is one of the most suitable species for aquaculture in Africa. Since the 1970's, it has been considered to hold a great promise for fish farming in Africa. Some other merits of African catfish are: high growth rate reaching market size of 1 kg in 5–6 months under intensive management conditions: highly adaptable and resistant to handling and stress; can be artificially propagated by induced spawning techniques for reliable mass supply of fingerlings; commands a very high commercial value where it is highly cherished as food in Nigerian homes and hotels (Olaleye, 2005.).

The Clariid freshwater fishes belong to the family Clariidae with a wide geographical distribution in Africa consisting of 14 genera (Teugels, 1986a) and 32 species (Teugels, 1986b) in Nigeria. Syndenham (1980) reported that the family consists of 5 subgenera namely: Clarias, Clarioides, Anguilloclarias, Platycephaceloides, and Brevicephaloides. *C.*

---

* Corresponding Author

gariepinus is the species native to Africa where it is grown although mostly on a subsistence level for food. The fish is hardy and adaptable to diverse environments even with poor water quality with its air breathing ability (Hecht *et al.*, 1996). *C. gariepinus* is a typically non-aggressive stalking predatory omnivore that hunts at night using non-visual primary sense organs especially the senses of touch through the barbels and tactile organs on the mouth and skin (Bruton, 1996).

The availability of gametes throughout the year is important to ensure a constant supply of fish. In captivity (25⁰C; 12h light per day), *C. gariepinus* gametogenesis is continuous once sexual maturity is reached. However, whereas females can be stripped of eggs after treatment with pituitary extracts, spermatogenesis and male reproductive behavior do not take place spontaneously, even after hormonal therapy. To obtain spermatozoa it is necessary to kill male brood fish or surgically remove the testes. Storing batches of spermatozoa by cryopreservation would significantly improve the reproductive potential of male catfish. The procurement of reliable broodstock (of good genetic quality), fingerlings and as juvenile fish for stocking ponds and fish farms has been a major set back in the development of catfish culture in Nigeria. This is because these cultivable species are not easily obtained from the wild. The development of cryopreservation procedures for sperm of Clarias species will aid in the recovery of threatened and endangered species as well as in the genetic selection and maintenance of lines of selected stocks. Cryopreserved sperm can also benefit commercial aquaculture industry by allowing females to be spawned when males are not available, decreasing the need to hold captive male as broodstock.

1.1 Success of cryopreservation in African catfish in Nigeria

Cryopreservation of fish spermatozoa has been the subject of many investigations. Successful cryopreservation depends not only on the right choice of cryoprotectant and extender, but also on the freezing protocol used. Cryoprotectant and freezing rate together determine the damage to spermatozoa due to intracellular ice crystallization.

The first two years of the NACGRAB- OAU Department of Animal Sciences cryopreservation project (2005-2007) were dedicated to optimization of cryopreservation protocols of the catfish sperm under short-term condition in deep freezer at -10 to -30°C (Oyeleye and Omitogun, 2007) and testing the viability of cryopreserved sperm by studying the ability of these cryopreserved sperm in fertilizing freshly spawned eggs (Omitogun *et al.*, 2006).

The second phase of the project (2008-2010) was dedicated to cryopreservation of the catfish sperm under long term conditions in liquid Nitrogen (-290⁰C) and testing the motility and ability to fertilize eggs. Evaluation of optimization and economic feasibility of cryopreserved sperm was also carried out (Omitogun *et al.*, 2010). To this end, cryopreserved sperm in liquid nitrogen in Dewar container was further diluted and cryopreserved from 3-8 months, then was taken to an identified and willing commercial catfish farm with the objective of testing the ability of cryopreserved sperm of African giant catfish to fertilize a whole clutch of eggs from a mature female catfish, normally used by commercial farmers and consequently confirm the viability of using cryopreserved sperm in normal commercial hatchery operations. Our hypothesis was that if cryopreseved sperm is practically tested on commercial scale and is proven economically feasible, being a true reflection of what was obtained in the laboratory (Oyeleye and Omitogun, 2007) then this will help to conserve male brood stock (Omitogun et al., 2010) which are normally slaughtered for fry production of catfish, and likewise ensure all-year round artificial propagation, helping the fish farmers in overcoming the problem of scarcity of male catfish breeders which are often encountered in the dry season.

1.2 Background information: Sperm: Egg ratio for optimum fertilization of catfish eggs

Cryopreservation of African catfish semen in liquid Nitrogen (LN_2) will invariably help us to conserve the genetic resources of our desirable male fish breeders for all year round artificial propagation and also help in overcoming the problems of scarcity of desirable male catfish breeders often encountered by the farmers most especially in the dry season and to meet high demand for catfish consumption (Oyeleye and Omitogun, 2007).

Sperm collection in African catfish as mentioned requires killing the male fish in order to excise the testes, it is important to maximize the use of a single male by optimization of sperm: egg insemination ratio. For fresh spermatozoa, the effective insemination ratio was estimated as 245 x 10^3 spermatozoa per egg in C. gariepinus (Steyn, 1987) and 50 x 10^3 spermatozoa per egg in Heterobranchus longifilis (Otenne et al., 1996.). Because a percentage of spermatozoa die during freezing and thawing processes, the effective insemination ratio for frozen spermatozoa should be higher. In channel catfish, 50 x 10^6 frozen-thawed spermatozoa per 0.5 ml straw enabled fertilization of 250 eggs (200 x 10^3 spermatozoa per egg; Tiersch et al., 1994). In blue catfish, Ictalurus furcatus, a minimum of 13,000 x 10^3 frozen-thawed spermatozoa per egg were needed to achieve 54% of control fertilization. In C. gariepinus, 49 x 10^3 live frozen-thawed spermatozoa per egg achieved a hatching rate (51.2%) equal to the control (51%).The insemination ratio was within the range 6 to 24 x 10^3 spermatozoa per egg. (Steyn, 1993). During ovulation the belly of the female will swell considerably due to water absorption of the ovary. The speed of the ripening process is dependent upon water temperature and likewise, the development process from fertilized egg to hatching is dependent upon water temperature (Coppens, 2009).

African catfish spermatozoa were first successfully cryopreserved by Steyn et al. 1985 who obtained 40% motility 24h after storage in LN_2. Glucose in combination with glycerol has been most widely used cryoprotective solution. Recently, glucose in combination with DMSO was also shown to be effective (Urbanyi et al., 1999). Freezing rates can be rapid (e.g., pellet freezing on dry ice or in LN_2 vapor) or slow (e.g., at fixed rates in programmable freezer (Steyn, 1993). In most cases, sperm quality was only evaluated in terms of motility after thawing. When fertilization was included in the evaluation, sperm: egg ratio was not optimized and was often excessive (Padhi et al.,1995). Using excess spermatozoa for fertilization obviously masks the quality of cryopreserved spermatozoa, making comparison of protocols difficult (Viveiros et al., 2000).

Methods for cryopreserving spermatozoa and optimizing sperm: egg dilution ratio in African catfish Clarias gariepinus was first developed by Viveiros et al., 2000) where 5 to 25% DMSO and methanol were tested as cryoprotectant, by diluting semen in Ginzburg fish ringer and freezing in 1-milliliter cryovials in a programmable freezer. To avoid an excess of spermatozoa per egg, post-thaw semen was diluted 1:20, 1:200 or 1:2000 before fertilization. Even frozen- thawed spermatozoa with low numbers of live cells yielded adequate hatching rates. They found out that the maximum sperm dilution ratio to achieve hatching rates similar to control was 1:200 without losing fertilization ability. However, at 1:2000 the hatching rates produced with frozen spermatozoa were lower than the control African catfish. Similarly *Heterobranchus longifilis* spermatozoa were diluted 1:3 before freezing and 1:10 after thawing and had the same fertilization ability (78.9%) as the control (81.1%). On the contrary for *Cyprinus carpio*, no spermatozoa survived when diluted higher than 1: 5 before and after freezing (Lubzens *et al.*, 1997).

Cryopreservation of catfish spermatozoa is useful as a routine method of gamete storage and management. However, the economic factor should also be considered. The technology of cryopreservation with the use of liquid nitrogen though desirable but is cost intensive. Therefore there is a need to study how the cryopreserved semen will be maximally utilized with good fertilizing results at the same time cost-effective and affordable for the farmers

To avoid wastage of cryopreserved spermatozoa per egg clutch after dilution with physiological salt solution, fertilization of various measures of egg clutches were tested in the present research with differently cryopreserved spermatozoa for optimization and for cost evaluation. In a second study the concentration of the semen was reduced, *i.e.*, diluted at a dilution ratio of 1:20 and 1:200 to verify the spermatozoa are not in excess and consequently be wasted.

Another study was carried out in order to assist the farmers to determine the approximate amount of egg clutch that will be adequate for a milliliter (ml) of cryopreserved semen without wasting spermatozoa in order to evaluate economic cost and profitability. The aim of this study was to verify the possibility of cryopreserving African catfish under long-term condition in liquid Nitrogen (LN_2) and evaluate the viability and fertility optimization of a specific amount (*i.e.*, 1 ml) of cryopreserved semen of African catfish cryopreserved in LN_2 using various cryoprotective agents with different measures of egg clutches. This paper aimed to establish a standard fertility ratio between a ml of semen and clutch of eggs in order to prevent wastage of semen; be able to maximize the resources and evaluate profitability of cryopreservation in liquid nitrogen by evaluating the effects on the cryopreserved semen as to motility and hatchability, the ability to hatch the eggs from a gravid female and survival of ensuing larvae.

2. Materials and methods

For the benefit of prospective users of cryopreservation of African catfish semen, the whole process is pictorially presented in Figures 1-12.

2.1 Husbandry of the broodstock

The broodstock of the African catfish, *C. gariepinus* were obtained from reliable farms in Ile-Ife and Ibadan, Nigeria and were transported in 25 litre tank opened at the top to the Wet Laboratory of the Department of Animal Sciences, Obafemi Awolowo University, Ile Ife.

The matured male and female broodstock were kept at constant temperature of 27°C in a 1000 litre tank connected to a source of water by a pipe connected from the reservoir plastic tank placed in an elevated stand in the laboratory and its drainage was located at the bottom of the tank for easy flow by gravity.

The broodstock were fed with an imported floating palletized feed i.e., CoppensᴿR feed (42% protein, ISO -170 certified, Netherlands) containing a large percentage of high quality fish meal, which is especially important to facilitate repeated spawning at a maintenance level of 1.5% body weight on a daily basis by gradual hand broadcast. The water quality was regulated through proper monitoring and replaced weekly.

2.2 Selection of broodstock

The sexually matured female was selected according to their swollen, reddish genital papilla and a well distended, swollen soft abdomen. A slight pressure was applied on the abdomen towards the genital papilla after which ripe eggs oozed out which were green-brownish in color and ripe eggs are generally uniform in size. The female broodstock was stocked in the hatchery for about 2 days without feeding so that the alimentary tract was empty at the time of stripping. It is very important that the collected eggs did not get contaminated. Sexually matured male broodstock was selected based on a reddish or pinkish pointed and vascularized genital papilla. The temperature of the broodstock kept in the tank was maintained at 25–27°C.

2.3 Preparation of extender-cryoprotectant

Two extenders were used in this study, phosphate buffered saline (PBS) and Ginzburg Fish Ringer (GFR) with pH of 7.4 and 7.6 respectively. The extenders were prepared as shown in Table 1 with Calcium-free Hanks Balanced Salt Solution (Ca-FHBSS) used in experiment 2. 14; after which they were sterilized for 20 min at 15 lbs/inch^2 using a pressure cooker to avoid contamination and deterioration of the spermatozoa.

2.4 Semen collection

A good quantity of the sperm of African catfish cannot be stripped and sperm can only be obtained after sacrificing it. Sexually matured male weighing 0.8± 0.2kg were selected and kept in a different tank of about 50 l capacity for about 18 h prior to the time of sperm collection .The male broodstock was dried with clean towel and then made unconscious by breaking its backbone. The body cavity was carefully opened with a pair of sterilized scissors without damaging the testes after which the two testes were dissected out. It was then removed with a pair of forceps, the blood veins cleared out and rinsed in saline solution. The testes were lacerated with a new and sharp razor blade; the milt was gently squeezed out and collected in a sterilized Petri dish. The whole process was carried out in a disinfected environment to avoid bacterial contamination which can lead to degradation of samples, transfer of pathogens and inaccurate estimation of motility. Sterilized instrument and aseptic techniques for collection of sperm was incorporated to reduce the contamination by bacteria. The volume of the extracted sperm was measured with a 5.0 ml sterilized syringe.

2.5 Cryopreservation of sperm, cryoprotective agents used

Different combination of cryoprotective (CPA) agents are shown in Table 2 as CPA-DP (DMSO+PBS), CPA-DF (DMSO+GFR), CPA-GP (Glycerol+PBS), CPA-GF (Glycerol+GFR), CPA-DGP (DMSO+Glucose+PBS) and CPA-DGF (DMSO+Glucose+GFR). Before cryopreservation of semen, motility of the fresh semen was evaluated in two trials of dilution: 1:1 and 1: 20. In these trials two different extenders were used: Phosphate buffered saline (PBS) and Ginzburg Fish Ringer (GFR). The volume of the extracted semen was measured with a 5.0 ml syringe (DISCARDM(R) NIG) which was diluted with the extender PBS and GFR for first trial on a ratio 1:1 v/v and 1:10 v/v respectively, then mixed evenly with differently prepared cryoprotective agent combinations at a ratio 1:1 (PBS) for fertility and hatchability evaluation.

In the second trial, the two extenders were used but at different sperm dilution ratio for PBS and GFR at a ratio 1:20 for both extenders. The resulting semen-cryoprotective agent solution in each trial after thorough mixing was dispensed into labeled 1ml cryotubes with a 2-step freezing protocol of first freezing on the chilled water blocks at -10^0 C for 10 min before it was finally transferred into the liquid nitrogen for a long-term preservation.

Composition (g/1000 ml)	Phosphate buffer saline (PBS)	Ginzburg Fish Ringer	Calcium-Free HBSS 200mOsmol/kg
NaCl	8.0	7.0	5.26
KCl	0.02	0.28	0.26
$CaCl_2$	-	0.33	-
$NaHCO_3$	0.23	-	0.33
Na_2HPO_4	1.15	-	0.04
KH_2PO_4	0.20	-	0.04
$Mg SO_4 7H_2O$	-	-	0.13
$C_6H_{12}O_6$	-	-	0.66

Table 1. Composition of the extenders (g/l) tested for cryopreservation of catfish sperm in liquid Nitrogen.

Cryoprotective agent (CPA)	DMSO	Glycerol	PBS	GFR	Glucose/ Sucrose
DP	10	-	90	-	-
DF	10	-	-	90	-
GP	-	10	90	-	-
GF	-	10	-	90	-
DGP	10	-	85	-	5
DGF	10	-	-	85	5

DMSO =Dimethylsulphoxide, PBS =Phosphate buffered saline, GFR=Ginzburg Fish Ringer.: DP = DMSO and PBS; DF = DMSO and GFR; GP = Glycerol + PBS; GF = Glycerol + GFR; DGP = DMSO + PBS + Glucose/Sucrose; DGF = DMSO + GFR + Glucose/Sucrose

Table 2. Composition (%) of the cryoprotective agents used

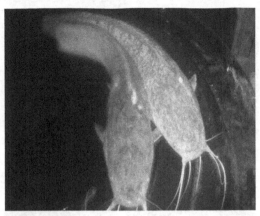

Fig. 1. The African catfish broodstock.

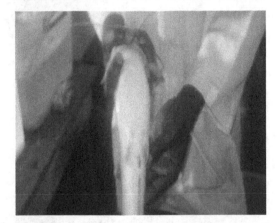

Fig. 2. Male catfish showing the genital openings.

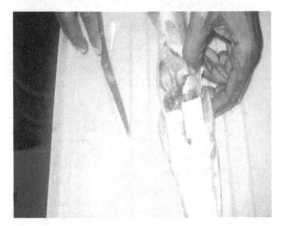

Fig. 3. Removal of testes from male catfish

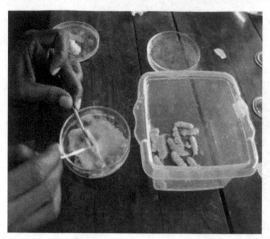

Fig. 4. Laceration of testes to extract milt

Fig. 5. Motility evaluation using a microscope and haemacytometer.

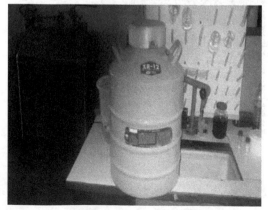

Fig. 6. Sperm cryopreserved in liquid Nitrogen for 4-8 months

Fig. 7. Removing the cryopreserved sperm and motility evaluated

Fig. 8. Rapid thawing process is employed.

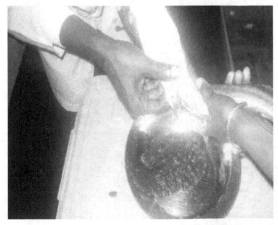

Fig. 9. Stripping the eggs from gravid female catfish after injection with Ovaprim

Fig. 10. Eggs divided into various clutch weights in Petri dishes

Fig. 11. Fertilized eggs incubated in aerated plastic containers covered with nets

Fig. 12. The temperature was kept at 25-27° C and sometimes covered with black plastic sheets.

2.6 Induced spawning and stripping

The readiness of the female broodstock to be used for breeding was tested by holding it in a head-up vertical position and a slight pressure was applied by pressing its abdomen with a thumb from the pectoral fin towards the genital papilla after which eggs run out freely. The selected broodstock were kept separately in different tanks without feeding them, after they were injected with 0.35 ml Ovaprim ® (Syndel, Canada) per kg live weight (Oyeleye and Omitogun, 2007) and then left for 10-12 hours latency period as a post ovulatory maturation period and to ensure high hatching rates and low proportion of deformed larvae (Hogendoorn, 1979).

2.7 Method of female stripping

The female body surface was gently dried with clean towel. It was tightly held at both ends by two persons with wet towels and stripped by a gentle press on the abdomen with a thumb towards the rear. The first free running eggs obtained at a slight pressing of the induced female broodstock were collected for fertilization (Legendre and Oteme, 1995).

2.8 Egg clutch variation and fertilization

After inducing, the female fish was stripped and the clutch of eggs weighed (about 150–160g/kg of the body weight). The eggs were weighed in various measures of 1.0g, 2.0g, 3.0g, 4.0g and 5.0g based on the level of each experiment. That is for 1.0g, it was weighed seven times with replicates for the different cryoprotective agents and control. After which it was fertilized with the cryopreserved semen in liquid nitrogen after thawing in warm water at 35^{0}C for 5 min. Fresh semen was used to fertilize same amount of clutches of eggs to serve as the control for both trials.

2.9 Motility evaluation

The motility of the spermatozoa before and after the addition of the cryopreservative agents, CPA and after thawing was evaluated for each trial. The cryopreserved semen was also further evaluated for fertility, hatchability and survivability for each trial. The motility test was done by diluting a drop of post thawed or fresh spermatozoa either with PBS, GFR or 0.9 % saline solution at a ratio 1:100 from which one drop of the solution was put on the hemocytometer and viewed subsequently under the microscope 10X and 40X, low and high power objectives of the microscope. The result arrived at is converted to the total number of spermatozoa per ml by multiplying it by the dilution factor (100) and 10^4 (SIGMA, 1994) as follows:

$$Total\ no\ of\ spermatozoa\ per\ ml = Average\ No.\ of\ counted\ spermatozoa\ x\ 10^4$$

$$of\ the\ cryopreserved\ semen \tag{1}$$

2.10 Fertility and hatchability evaluation

The development process from fertilized eggs to hatching is dependent upon water temperature while hatching rate is, next to egg quality, dependent on the water quality;

temperature, oxygen level, pH and water hardness. After stripping of the induced female broodstock, the eggs were weighed in grams depending on the on-going experiment *i.e*, 1.0g (600±100 eggs), 2.0g, etc. The various measures of eggs (repeated 12 times together with replicates) were fertilized with cryopreserved semen thawed at 35⁰C for 5 minutes, and a pair of egg clutches with fresh semen as control experiment. The mixture of eggs and semen was stirred gently for at least 1.0 min to allow contact and adequate fertilization. Within a few minutes after fertilization the eggs absorbed water and could become sticky so the eggs were distributed in a netted basket suspended in the hatching trough (50cm x 35cm x 30cm) containing contaminant-free (passed through a purification system with ultraviolet sterilization at 3000 µW/cm²) well-aerated water in a single layer so that the eggs get sufficient oxygen during incubation. The hatching troughs were completely covered with mosquito net and black polythene materials placed under 200 Watt bulbs to prevent mosquitoes and other insects laying eggs and to increase level of heat generation. The system was supplied with an electric aerator to increase level of oxygen dissolved in the water.

The incubated eggs were monitored and temperature maintained between 26⁰C -27⁰C for incubation between 23– 25 h. Soon after hatching the larvae passed through the net and the dead eggs and shells remain on the net in the basket. The larvae were then simply separated from the unfertilized eggs and eggshells by lifting the basket and the nets out of the hatching trough.

The percentage, % fertility and hatchability were determined subjectively after 12–15 h of fertilization by identifying the healthy developing eggs which were transparent green brownish in colour (Coppens, 2007) while the dead eggs were also estimated:

$$\% \ Fertility = (No. \ of \ fertilized \ eggs \ / \ No. \ of \ inseminated \ egg) \ X \ 100\% \tag{2}$$

$$\% \ Hatchability = \frac{Total \ No. \ of \ fertilized \ eggs \ - No. \ of \ unhatcched \ eggs}{Total \ No. \ of \ Fertilized \ eggs} \times 100\% \tag{3}$$

2.11 Post-hatching survivability evaluation

This is done by allowing the newly hatched larvae of all the treatments and that of the control to live on the remains of their yolk sacs for the first 4 days (Heicht *et al.*,1996) after hatching out of the eggs and thereafter carefully removed from the hatching troughs and were fed with Artemia (Inve Aquaculture, USA) on a regular basis (*i.e.,* twice per day). Irregularities in the activities of the fry in terms of feeding, movement in water was observed at the same time taking note of the dead fry which were removed immediately to avoid contamination of water. Survivability evaluation which was observed for a period of about 3 – 4 weeks was done for each stage of the experiment together with fertility and hatchability for fresh (control experiment) and cryopreserved spermatozoa. The post-hatching survivability was evaluated as follows:

$$\% \ Survivability = \frac{Total \ No. \ of \ larvae \ - \ No. \ of \ dead \ larvae}{Total \ No. \ of \ larvae} \times 100\% \tag{4}$$

.12 The control

The control for both trials was prepared by the use of fresh semen obtained from the acerated testes from a normal gravid male broodstock with the use of sterilized dissecting knives but activated with saline solution in the ratio of 1:1 v/v and subsequently used to ertilized various measures of egg clutches ranging from 1.0g, 2.0g, 3.0g, 4.0g and 5.0g normal eggs from the same batch of eggs i.e. from the same fish. Control was set up for the evaluation of each parameter for trials *i.e.*, motility, fertility, hatchability and survivability.

The motility evaluation of the post-thawed cryopreserved sperm was evaluated after dilution with the extender on a ratio 1:100 using hemocytometer (SIGMA, 1994).

2.13 Further scaling up for commercial application of cryopreserved semen dilution and egg clutch fertilization

Motility of the fresh semen was evaluated in two different trials of dilution, before any cryopreservation of the semen. Phosphate-buffered-saline (PBS) and ordinary saline water were the two extenders used in the two different trials and were diluted 1:1 and 1:40. The extracted semen volume was measured with a 5.0ml syringe (DISCARDIM(R)NIG) which was diluted with the extender PBS for trial on a ratio 1:20v/v and 1:200v/v respectively and thereafter mixed evenly with the cryoprotective agents at a ratio of 1:1 and cryopreserved for the next seven months in liquid Nitrogen stored in Dewar container

For the second trials, the same cryoprotectant (85% PBS+5% glucose+10% DMSO) was used but at different sperm dilution ratio of 1:40. In both trials, the resulting semen in each experiment after thorough mixing was then dispensed into labeled 1ml cryotubes while a 2-step freezing protocol, e.g. initial freezing onto frozen water (ice) blocks at -10^{0}C for 30min before the final transfer into liquid nitrogen for the next 4 to 7 months.

After induction with Ovaprim, the female broodstock was stripped and the clutch of eggs weighed (which was about 150-200g/kg body weight) and divided into three portions of about 120 g each for the experiment on fertilization with cryopreserved sperm of diluted 1:1 and 1: 40 and for the control.

This procedure was repeated three times in a nearby commercial farm 5kms away from the University to serve as the replicates and to ensure the repeatability of the experiments.

2.14 Experiment on storing sperm in refrigerator

The various extenders used were

- 200mOsmol/kg Ca-F HBSS
- 300 mOsmol/kg Ca-F HBSS
- 400 mOsmol/kg Ca-F HBSS
- RPMI 1640 (SIGMA) Culture Medium in 0.9% NaCl Solution

The three different osmolalities of extender Ca-FHBSS (Calcium-free Hanks Balanced Salt Solution) were prepared according to Riley, 2002. The sperm with the 200, 300 and 400 mOsmol/kg Ca-F HBSS were kept in the refrigerator at 4^{0}C. The semen samples with RPMI and 0.9% NaCl solutions were kept at both room temperature and refrigerator. Two replicates were made for each treatment.

2.15 Statistical analysis

The data collected on the parameters, motility, fertility and hatchability was subjected to standard statistical analysis. The data collected were analyzed using analysis of variance (ANOVA) to find a level of significance at $p < 0.05$. The number of motile sperm counted per square of hemocytometer and percentage of motile sperm obtained after cryopreservation were subjected to Duncan's multiple range test to evaluate effects of types of cryopreservatives as well as egg clutch weights and sperm dilutions and period of cryopreservation on sperm motility. The data collected on motility, fertility and hatchability in the first trial were subjected to 2- way ANOVA at a significance level of $p < 0.05$. The bar charts and line chart showing the relationship between the period of refrigeration and the % motility for the various extenders were also employed for better understanding of the results.

2.16 Cost of production of cryopreserved semen

The costs of chemicals and other consumables used for the study were listed in Table 5. The rates per gram or per ml were calculated to determine the effective cost per ml of the cryopreserved sperm (Table 6).

3. Results and discussion

3.1 Effects of different cryoprotectants on fertility, hatchability, motility and survivability

In the first trial with dilution ratio of 1:1 (sperm: extender) the effect of nature of cryoprotectants on the parameters measured was significant ($p<0.05$). Though dimethylsulphoxide+5% glucose+ PBS (DGP) was higher, DGP and DP gave the best results and was not significantly different ($p>0.05$) from each other but significantly different ($p<0.05$) from other cryoprotectants. It was followed by GP but not significantly different ($p>0.05$) from other cryoprotectants. Fertility also followed the same trend, but DGP was significantly different from DP from their LSD values followed by GP. GF gave the least result but significantly different ($p<0.05$) from other cryoprotectants. Related trend was also observed for other cryoprotectants but that of control was higher ($p<0.05$) than other treatments for each parameter. A similar trend was also observed for motility and survivability.

In Table 3, control (fresh semen), DGP and DP were compared. It is obvious that control has highest mean value (significantly different), this may be expected as the control was not passing through any treatments and processes.

In the second trial when the semen was diluted at ratio 1:20 (sperm: extender) a very close trend was observed which shows that the spermatozoa seems to be too much and probably wasted in the first trial. Besides control which was significantly different ($p<0.05$) for all the parameters, DGP gave the best result before DP, this may be explained by extracellular protection offered by the glucose, but the mean values are not significantly different ($p>0.05$) from each other. Fertility also related to hatchability followed the same trend with hatchability. In fertility, DP, DGP and GP were not significantly different from each other. However, the mean values are significantly different from each other for other parameters in decreasing order of C > DGP > DP > GP > GF. Generally, DGP gave the best followed by DP (without 5% glucose) while Glycerol in combination with Ginsburg fish ringer gave the lowest result.

A close means values were also discovered with trial 1 which indicate the results were better in trial 2, the further diluted semen which could supposedly be explained by addition of extenders.

The differences in fertility and hatchability with control and cryoprotectants tested may be due to the mild damage done to the spermatozoa during the process of lacerating the testes to extract the semen, and also due to the intracellular vitrification (Cryobiosystems, 2009)- a commonly occurring problem in the process of cryopreservation in liquid Nitrogen.

3.2 Effects of egg clutch weight on viability of African catfish gametes

There was a significant effect ($p < 0.05$) of egg clutch weight on fertility, hatchability and survivability. In trial 1, although, the hatchability increases with increase in egg weight but the increment at egg weight 4.0g and 5.0g was not significantly different (LSD=88.749, $p > 0.05$). There was fertility optimization at 4.0 g of egg clutch weight which though, close to the mean values of 5.0g which is higher but not a uniform increase. The same trend was observed for fertility of eggs but much higher than corresponding hatchability which may be due to loss of eggs to external factors like temperature, contamination and possible error during record taking. There is no significant difference for fertility at egg clutch weight 3.0g and 4.0g but there was significant difference ($p < 0.05$) from 5.0g. Survivability was not significantly different from each other except for 5.0g ($p < 0.05$).

In the second trial, a similar trend was observed; hatchability was highest ($p < 0.05$) at egg clutch weight 4.0g. No significant difference ($p > 0.05$) in survivability was observed except for egg weights 3.0g and 4.0g.

Generally, for both trials, egg clutch weight at 4.0 g gave the optimum viability value.

3.3 Effect of type of cryoprotectant and egg clutch weight interaction on hatchability and fertility

From the statistical analysis, it showed that there could also be effect of interaction of both cryoprotective agent (CPA) and egg clutch weight on fertility and hatchability.

The result, as observed shows a significant effect of ($p < 0.05$) of interaction of CPA and egg clutch weight on fertility, hatchability for the first trial and only on fertility for the second trial.

Effect of interaction of cryoprotectant on hatchability was not different from the trend of results obtained in previous results. However, the cryoprotective agents were not significantly different ($p > 0.05$) from each other but DGP, GP and DP still maintained the higher mean values while control took the highest. DP and DGP were not significantly different ($p > 0.05$) from each other for egg clutch weight such as 1.0g to 5.0g. However, GP was significantly different ($P < 0.05$) from DGP and DP for egg clutch weight 2.0-5.0g. For GF, DF and DGF, there was also no significant effect ($p > 0.05$) with changes in egg weights. The effect of interaction of both CPA and egg weights on fertility was also significant ($p < 0.05$).

The results for both trials were also very close, following the same trend, except for control changes in egg clutch weights from 1.0g to 5.0g was generally not significantly different for DP and DGP followed by GP and GF which were not significantly different (p>0.05) from means values of egg clutch weight 4.0g–5.0g for DGP and DP. However, GF gave the lowest value for the two trials.

3.4 Scaling up of the applications of the cryopreserved semen for commercial aquaculture

3.4.1 Effect of dilution ratio on viability of catfish gametes

Table 3 shows the comparison for the parameters measured among the control, dilution ratios 1:1 and 1:40. The fresh semen gave the highest fertility and hatchability rates (P<0.05). It is significantly different from ratio 1:1 and ratio 1:40. Comparing the fertility and hatchability rates of the two different dilution rates, ratio 1:1 gave the highest fertility and hatchability rates which was significant (P<0.05). Survival rate however, followed a different trend in which dilution ratio 1:1 gave the highest survival rate closely followed by ratio 1:40 while the control semen gave the least survival rate (P< 0.05).

Most importantly, the differences in fertility and hatchability may also be attributed to the condition of the farmer's hatchery environment in which many environmental and sanitation conditions were compromised for maximum profit (Amupitan *et al.*, 2010).

DILUTION RATIO	MOTILITY	FERTILITY	HATCHABILITY	SURVIVAL
1:1	55%[b]	30% [b]	35%[b]	15%[a]
1:40	49%[b]	29%[c]	34%[c]	14%[b]
C	72%[a]	54%[a]	62%[a]	13%[c]
LSD	13%	9.2%	25%	13%

Means in the same column with different letter are significantly different at P<0.05, C=Control, LSD = Least Significant Difference

Table 3. Effects of dilution ratio on viability of Catfish semen diluted at ratio 1:1 and ratio 1:40

3.4.2 Effect of dilution ratio on motility of catfish semen and survival of ensuing larvae

The fresh semen gave the highest motility at appreciably high percentage (71.00%) which was significantly different (P<0.05) from cryopreserved semen diluted at ratios 1:1 (50.52%) and 1:40 (49.05%) (Fig. 13). However, there was no significant difference between the two diluted cryopreserved semen (P>0.05). It was evident that the freezing process and cryopreservation decreased sperm motility after cryopreservation. It could be deduced that cryopreserved sperm still needs to be completely activated after thawing in order to fertilize the whole clutch of eggs since there is a direct relationship between motility and fertility.

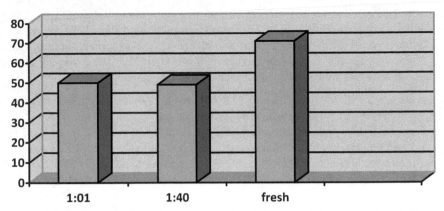

Fig. 13. Motilities of cryopreserved semen at ratios (1:1, and 1:40) compared with the control fresh semen

Cryopreservation in liquid Nitrogen did not have any effect on the survival of *C. gariepinus* larvae produced from cryopreserved semen as shown. However, larvae produced by the cryopreserved semen gave a higher survival, ratio 1:1 (65%), Ratio 1:40 (63%) and fresh semen (50%) (P>0.05).

The present study proves that sperm of African catfish cryopreserved aged up to 7 months in liquid Nitrogen and diluted more than 40 times with the extender is viable. The reason for low survivability rate in the control experiment using fresh semen may be attributed to high stocking density (because of the greater number of surviving fry) as practiced by many farmers, *i.e* the quantity of larvae per unit volume of water is less for cryopreserved sperm which in turn was favourable for survival.

3.5 Motility evaluation of refrigerated catfish sperm cells in different extenders

This experiment evaluated the effect of extenders and period of refrigerated storage on the sperm motility of *Clarias gariepinus* sperm cells with the intent to identify a suitable extender for the refrigerated storage of the sperm cells of *Clarias gariepinus*. Semen samples were collected from mature broodstock and were refrigerated with various different extenders at ratio 1:3 namely: Calcium-free Hanks' Balanced Salt Solution (Ca-F HBSS), RPMI 1640 culture medium and 0.9% NaCl. Ca-F HBSS extender was prepared at 3 different osmolalities: 200mOsmol/kg, 300mOsmol/kg and 400mOsmol/kg. Sperm in RPMI 1640 and 0.9% NaCl extenders were also kept at room temperature to assess the effect of refrigeration on motility of catfish sperm cells. Motility was monitored on a 24-hour basis and % motility was evaluated daily. Results showed that sperm cells of *Clarias gariepinus* using 200mOsmol/kg as extender (p<0.05) can be stored under refrigeration for 12 days. However, of all the extenders evaluated, RPMI 1640 proved to be the most effective extender (p<0.05) retaining higher motility of the refrigerated sperm cells of *Clarias gariepinus*.

3.5.1 Effect of refrigeration on % motility of sperm cells

The semen samples extended with 0.9% NaCl and the RPMI culture solution at room temperature did not have motile sperm cells after 48 hours. The motility of the semen sample with 0.9% NaCl at room temperature dropped from the initial motility of 74.82% to 6.24% (Fig.3.6. 1) after the first 24 h while the semen sample with the RPMI culture solution at room temperature had 0.4% motility at the end of 24 h. The semen sample with 0.9% NaCl at room temperature had a fishy irritating smell after 24 h. This may be due to the production of waste since the sperm cells metabolised at the normal rate.

The semen samples with 0.9% NaCl and RPMI 1640 culture solution retained motility much longer when refrigerated (Fig. 14). The refrigerated semen sample with 0.9% NaCl retained motility for up to 7 days with motility after 24, 48, 72, and 168 hourly being 34.75%, 17.03%, 14.06% and 4.46% respectively. The refrigerated semen sample with RPMI however kept for 9 days with motility after 24, 48, 72, 168 and 216 hours being 53.47%, 37.62%, 25.64%, 8.32% and 5.25% respectively.

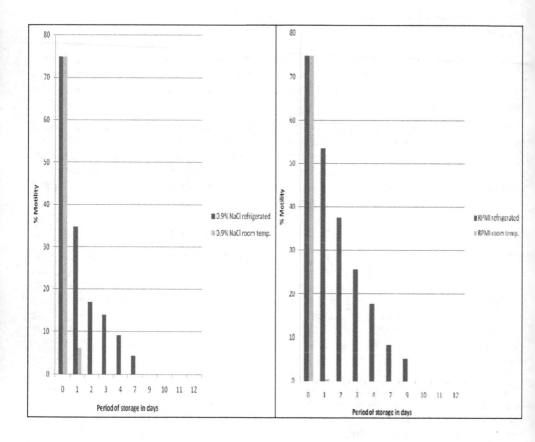

Fig. 14. Effect of refrigeration on motility of semen sample extended with NaCl (left) and RPMI (right). The refrigerated semen sample retained motility till the 7th (in NaCl) and 9th day (in RPMI) whereas semen samples at room temperature only lasted till the first day after storage.

The art of refrigeration provides a low temperature which lowers the metabolic rate of living organisms. The extended semen samples maintained at room temperature proceeded at the normal metabolic rate, hence the sperm cells could not survive up to 48 h. When the sperm cells were still within the fish in the testis, they were supplied with nutrients and the waste they produced are excreted out of the testis, they cannot be supplied with energy or nutrients except provided externally as in tissue culture. Their wastes also accumulate in the solution in which they are suspended in.

High metabolic rate means faster rate of using up available resources by living organism such as nutrients and energy. It also means that waste will be produced at a faster rate thereby causing fast accumulation of waste in the solution in which the sperm cells are suspended. This accumulation will immediately reach a toxic level causing fatality in the sperm cells. Whereas the low temperature provided by the refrigerator to the refrigerated semen sample reduced the metabolic rate of the sperm cells, thereby reducing the rate at which the available nutrients and energy in the semen-extender solution are used up. The nutrient and energy in the semen-extender solution lasted a much longer period when refrigerated, thus keeping the sperm cells alive for a longer period than in the semen samples extended at room temperature.

3.5.2 Effect of osmolality of Ca-F HBSS on % motility

After 24 hours, the motility of the refrigerated semen sample with the 200mOsmol/kg Ca-F HBSS dropped to 34.85% and motility was retained till the 12[th]day (288 hours) with 0.5% motility. The refrigerated semen sample with 300mOsmol/kg Ca-F HBSS retained motility for 10 days with % motility at 24 h being 31.88 and motility by the tenth day had dropped to 2.28%. The refrigerated semen sample with 400mOsmol/kg Ca-F HBSS retained motility also for 10 days but with a lower % motility at the 10th day (0.4% motility) but with motility at 24 h being 34.46% (Fig. 15).

Based on the length of days for which motility was retained, the 200mOsmol/kg Ca-F HBSS proved to be a good extender since it retained motility for 12 days but with a very low motility (0.5%) However the 300mOsmol/kg, although retained motility for only 10 days, is better since it had the highest motility at the fourth, seventh, ninth and tenth day (i.e. 16.93%, 10.73%, 2.77% and 2.28%).

A good extender should be isotonic to the seminal plasma of the fish. This is to keep the sperm cells immotile until ready for use. Sperm cells are immotile in the seminal plasma and when semen is released in aquatic environment, osmolality goes down and motility is initiated in freshwater species. (Maria et al., 2006). Motility in freshwater species is initiated by exposure of the semen to a hypotonic solution (Morisawa and Suzuki, 1980). Use of extender solutions that are similar in chemical concentration and osmolality are essential to optimizing storage time (Baynes et al., 1981). According to Mansour et al., 2002, motility of Clarias gariepinus is completely but irreversibly suppressed in electrolytes and non-electrolytes with an osmolality of 200mOsmol/kg. This statement by Mansour et al., 2002 proves that the osmolality of the seminal plasma of Clarias gariepinus is less than or equal to 200mOsmol/kg. This explains why the 200mOsmol/kg of Ca-F HBSS retained motility till the twelfth day as it is closer to being isotonic to the seminal plasma of Clarias gariepinus.

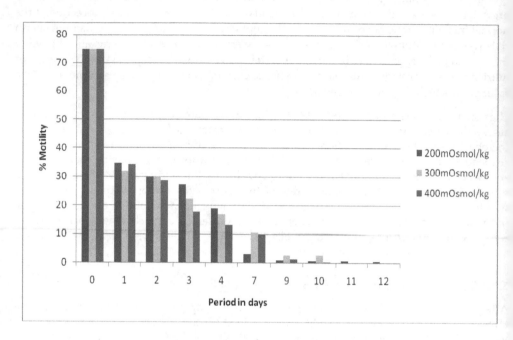

Fig. 15. Effect of osmolality on the sperm motility of refrigerated semen sample in Ca-F HBSS The 200mOsmol/kg Ca-F HBSS extended semen sample retained motility up to day 12 (0.5%)

3.5.3 Effect of extenders on % motility

The refrigerated semen sample with the RPMI culture solution had relatively very high motility at 24 hours after refrigeration; it had a motility of 53.47% as against the 34.85% motility of the semen sample with the 200mOsmol/kg Ca-F HBSS which comes next in rank with it after 24 h. The RPMI extended semen sample also had the highest motility after 48 h of refrigeration (37.62%) with the next in rank being semen sample extended with 300mOsmol/kg Ca-F HBSS with motility of 30.10%. However, the RPMI extended semen sample retained motility only till the fifth day with motility at the fifth day being 5.25% (Fig.16).

The control experiment being semen samples extended with 0.9% NaCl solution retained motility for 7 days with motility at the seventh day being 4.46% and its motility at 24 hours was 34.75% which is exceeded by the semen sample extended with the 200mOsmol/kg Ca-F HBSS which had 34.85% motility at 24 h (Fig. 17).

The relatively high motility retained by the refrigerated semen sample with RPMI culture medium may be due to the additional nutrient supply provided by the RPMI culture medium solution. The RPMI medium culture contains many amino-acids, vitamins and growth factors.

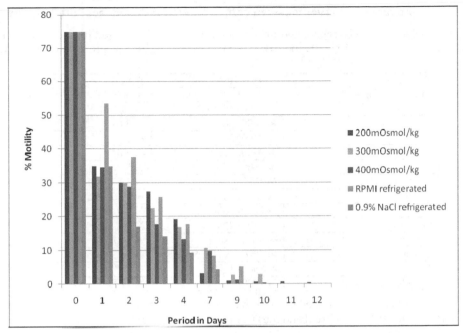

Fig. 16. Effect of the different extenders on the motility of the sperm cells of *Clarias gariepinus*.

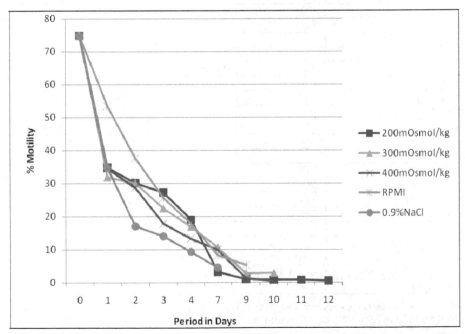

Fig. 17. The decline in motility as the period of storage of refrigerated sperm increased in different extenders.

3.6 Cost of production of cryopreserved catfish sperm for the aquaculture industry

The cost analyses for each reagent and materials used in cryopreservation per ml, g or piece is shown in Table 5. Table 6 shows the cost estimates for the 4-best cryoprotecting agents (DP,

ITEMS	COST (in Naira)	RATE
Liquid Nitrogen LN_2	₦16000/20 l Dewar	₦ 800/l
DMSO	₦ 3700/100/ml	₦ 37/ml
NaCl	₦ 1800/500g	₦ 3.6/g
KCl	₦ 2200/500g	₦ 4.4/g
Na_2HPO_4	₦ 3400/500g	₦ 6.8/g
KH_2PO_4	₦ 2600/500g	₦ 5.2/g
Distilled water	₦100/l	₦ 0.1/ml
$NaHCO_3$	₦ 6028/500g	₦ 12.056/g
$CaCl_2$	₦ 2200/500g	₦ 4.4/g
Glycerol	₦ 2500/2.5l	₦ 1.00/ml
Glucose	₦ 4,100/500g	₦ 8.2/g
Cover Slip	₦175/box (100pcs)	₦ 1.75/slip
Cryovials	₦ 8500/1000	₦8.50/tube
Broodstock (male)	₦ 1200	₦ 1200

Table 5. The cost of reagents and other consumables used to cryopreserve African catfish semen in liquid Nitrogen (LN_2) and listed and the rates per l or ml, g, or per piece are estimated

Ingredients	DP	DGP	GP	GF
Broodstock	1200	1200	1200	1200
Liquid Nitrogen	5000	5000	5000	5000
Glycerol	-	-	1.00	1.00
NaCl	5.76	5.76	5.76	5.76
KCl	0.0176	0.0176	0.0176	0.0176
Na_2HPO_4	1 .564	1 .564	1 .564	-
KH_2PO_4	0.208	0.208	0.208	-
$CaCl_2$	-	-	-	0.2904
$NaHCO_3$	-	-	-	0.5546
Distilled Water	50.00	50.00	50.00	50.00
Glucose	-	4.10	-	-
Cover slip	1.75	1.75	1.75	1.75
Cryovials (80 paces)	680.00	680.00	680.00	680.00
Saline Solution	3.24	3.24	3.24	3.24
Miscellaneous	300.00	300.00	300.00	300.00
TOTAL	7279.5396	7293.6396	7243.5396	7192.1214
Cost/unit of 80 cryovials	7279.5396	7293.6396	7243.5396	7192.1214
Cost per 1ml cryovial	₦ 90.9942	₦ 91, 0455	₦ 90.5442	₦ 89.9015
Proposed Selling price	₦ 100.00	₦ 100.00	₦ 100.00	₦ 100.00
Profit /cryovial	₦ 9.00	₦ 8.95	₦ 9.46	₦ 11.01

Estimation based on 200 ml extender for each cryoprotective agent: DP:DMSO-PBSS, DGP: DMSO-Glucose-PBS; GP: Glycerol-PBS; GF: Glycerol-Fish Ringer .

Table 6. Cost estimates/ml in Naira (₦) for preparation of cryopreserved African catfish semen using different cryoprotectants

DGP, GP and GF. Though, they were not all procured from the same source, all gave almost similar total cost which ranges between ₦7,192. 00 – ₦7,280. 00 and cost/vial ranges between ₦ 89.00 and ₦ 91.00 (1 $ = ₦ 150.00; ₦, naira is the Nigerian currency). Consequently, the proposed economical selling price (in case a cryobank in the University or the collaborating research agency is to be established) considering the present inflation rate is the same for the type of cryoprotectant to be used, but net profit per vial is slightly different.

The selling price for GF is expected to be reduced irrespective of its total cost because of lower viability rate of semen cryopreseved with it. Profit per vial was also calculated which is highest for GP at ₦100/vial and lowest for GF at ₦95/vial. The selling price ranges from ₦95.00 – ₦100.00 which would encourage the buyers to buy it affordably with reasonable profit for the institute (cryobank).

However, the total cost of GP and GF are lower because of the lower cost price of glycerol, but GF gave the least total cost because of a slight difference in the type and amount of chemical used. Total cost of DGP is higher than that of DP, because of cost of glucose inclusion and higher cost of DMSO.

4. Conclusion

From this study it can be concluded that at further dilution of semen together with different cryoprotecting agents, viability of the semen is still maintained but it varies depending on the type of cryoprotecting agents used. DMSO dimethylsulphoxide proved to be more efficient than other cryopreservatives in preserving sperm viability. Its potential in cryopreservation can be increased when used in combination with a 5% glucose solution, i.e. DGP and DP proved to be the best even for both trials.

Also, viability of African catfish C. gariepinus semen cryopreserved in liquid nitrogen can be maintained for a relatively long time provided ideal protocols are strictly followed. From economic feasibility perspective of cryopreservation of catfish semen, cryopreserved semen is economically feasible and profitable for the cryobank institute or company. The farmers are also assured of the viability of the sperm cells they are buying.

The ability of the African giant catfish (Clarias gariepinus Burchell, 1822) semen cryopreserved from 4-8 months with different combinations of extender and cryoprotecting agents, dimethylsulphoxide (DMSO) and glycerol with two extenders: GFR (Ginzburg fish ringer) and PBS (Phosphate buffer saline) to fertilize various egg clutch weights were investigated to evaluate the optimum clutch of egg a milliliter of cryopreserved semen can fertilize. DMSO + glucose with PBS and DMSO+ PBS only proved to be the best cryoprotectant-extender combination in maintaining viability of catfish semen. The optimum viability of the semen was also observed at 4.0-5.0 g of clutch of eggs/ml of semen with little deviation. The first trial was on dilution ratio of 1:1 but in the second trial, the semen was diluted further at a ratio of 1:20 and tested on various egg clutch weights (1, 2, 3, 4, 5 g) to evaluate the viability of cryopreservation even at further dilution. There was a significant effect of different cryoprotecting agents ($p<0.05$) on egg clutch weights. There was a significant effect ($p<0.05$) for hatchability and fertility in the first trial but only fertility in the second trial.

In another experiment we tried higher dilution ratio 1:40 and with bigger clutch of eggs (120g) of a standard female breeder (1.0 kg + 0.2 kg) simulating the practices of many commercial farmers, this time with the cryopreserved sperm. Compared with the control fresh semen that gave the highest motility at appreciably high percentage (71%) which was significantly different (P<0.05) from cryopreserved semen diluted at ratios 1:1 (50.52%) and 1:40 (49.05%). However, there was no significant difference between the two diluted cryopreserved semen (P>0.05). It was evident that the freezing process and cryopreservation decreased sperm motility after cryopreservation. It could be deduced that cryopreserved sperm still needs to be completely activated after thawing in order to fertilize the whole clutch of eggs since there is a direct relationship between motility and fertility.

In order to assist subsistence fish farmers who may not be able to obtain cryopreserved semen, the motility of sperm cells stored under refrigerated conditions in different extenders was studied. This research evaluated the effect of extenders and period of refrigerated storage on the sperm motility of *Clarias gariepinus* sperm cells with the intent to identify a suitable extender for the refrigerated storage of the sperm cells of *Clarias gariepinus*. Semen samples were collected from mature broodstock and were refrigerated with various different extenders at ratio 1:3 namely: Calcium-free Hanks' Balanced Salt Solution (Ca-F HBSS), RPMI 1640 culture medium and 0.9% NaCl. Ca-F HBSS extender was prepared at 3 different osmolalities: 200mOsmol/kg, 300mOsmol/kg and 400mOsmol/kg. Sperm cells in RPMI 1640 and 0.9% NaCl extenders were also kept at room temperature to assess the effect of refrigeration on motility of catfish sperm cells. Motility was monitored on a 24-hour basis and % motility was evaluated daily. Results showed that sperm cells of *Clarias gariepinus* using 200mOsmol/kg as extender (p<0.05) can be stored under refrigeration for 12 days. However, of all the extenders evaluated, RPMI 1640 proved to be the most effective extender (p<0.05) retaining higher motility of the refrigerated sperm cells of *Clarias gariepinus*.

The viability of sperm preserved under ordinary refrigerated conditions is possible for a short period of time of 2-7 days depending on the amount of extender used. A culture medium like RPMI 1640 used in this study may give longer life span for sperm cells under refrigerated conditions. Extenders like RPMI 1640 or the cheaper Ca-F HBSS is an alternative that can be recommended for farmers who may have excess of sperm cells from slaughtered male fish for more female gravid eggs that can be sourced within a time period of one week.

The cost of production of a cryotube of sperm (cost of materials, reagents, liquid nitrogen, etc.) was carried out to determine the cost of a milliliter of cryopreserved sperm with a view to selling cryopreserved semen by the research laboratory to farmers who may not be able to afford to buy a male broodstock yielding an affordable cost of ₦100/ml compared to current cost of a male breeder which is ₦ 1000 -1500 each (1 $ = 165 ₦). This also ensures the farmer that the cryopreserved sperm cells they might alternatively buy are viable and will be able to induce the spawning of the female broodstock.

5. Acknowledgements

This research was made possible by research grants from the OAU University Research Committee (11812 AVT), National Centre for Genetic Resources and Biotechnology, Ibadan and the World Bank-Assisted STEP-B (Science and Technology Post-Basic), Abuja. The authors are grateful for the technical assistance of Messrs. Akin Babatunde, Samuel Oladejo, Rotimi Solanke and Wasiu Olaniyi. Our sincere gratitude goes to Prof. B. Solomon of NABDA (National Biotechnology Development Agency) for partially funding the publication charges and Dr. Igor Katkov and Mrs O. Osoniyi for their comments and their expert editing of the manuscript.

. References

Amupitan, P.S, Ilori, O.I, Aladele, S.E, Odofin, W. T. and Omitogun, O.G. (2010). Scaling up the viability of cryopreserved giant catfish semen in liquid nitrogen for commercial application. In: *34th Annual Conference of the Genetic Society of Nigeria.* NIHORT, Ibadan. Pp 335-343. ISSN 0189-9686

Baynes, S.M., Scott, A. P. and Dawson, A. P. (1981). Rainbow trout spermatozoa: Effect of cations and pH on motility. *Journal of Fish Biology* Vol.19. pp. 259-267.

Bruton, M.N. (1996). Alternative life-history strategies of catfishes. In: Legendre, M. and J.P. Proteau (Eds.).The Biology and culture of Catfishes. *Aquatic Living Resources,* Paris Vol. 9 Hors Series, pp. 35-41.

Coppens International. (2009). www.coppens.eu

Cryobiosystems. (2006). Fundamentals of cryobiology-
http://www.Cryobiosystem.imv. com/CBS/Cryobiology/cons.cbs-asp.

Hetch,T., Oellermann, L. and Verheust, L.. (1996). Perspectives and clariid catfish culture in Africa. In: Legendre, M. and Proteau, J-P. (eds.).The Biology and Culture of Catfishes. *Aquatic Living Resources. Paris,* 1996, Vol. 9, Hors Series pp. 197-206.

Hogendoorn H. (1979). Controlled propagation of the African catfish, *Clarias lazera (C & V).*I. Reproductive biology and field experiments. *Aquaculture* Vol. 17 pp. 323-333.

Legendre, M. (1986). Seasonal changes in sexual maturity and fecundity, and HCG-induced breeding of the catfish *Heterobranchus longifilis* Val. (Clariidae) reared in Ebrien lagoon (Ivory Coast). *Aquaculture* Vol. 55 pp. 201-213.

Legendre, M. and Oteme, Z.I. (1995). Effect of varying latency period on the quality and quantity of ova after HCG-induced ovulation in the African catfish, *Heterobranchus longifilis* (Teleostei: Clariidae). *Aquatic Living Resources.* Paris Vol. 8 pp. 309-316.

Lubzens E., Daube, N., Pekarsky, I.., Magnus, Y., Cohen, A., Yusefovich, F. and Feigin, P. (1997). Carp (*Cyprinus carpio* L.). Spermatozoa cryobanks strategies in research and application. *Aquaculture* Vol. 155 pp. 173-178.

Mansour, N, Lahnsteiner, F. and Patzner, R. A. (2002). The spermatozoan of the African Catfish: fine structure, motility, viability and its behaviour in seminal vesicle secretion. *Journal of Fish Biology* Vol. 60 No. 3. pp. 545-560.

Maria A. N., Viveiros, A. T. M., Orfao L. H., Oliveira, A. V. and Morales, G. F. (2006). Effects of cooling and freezing on sperm motility of the endangered fish Piracanjuba *Brycon orbignyanus* (Characiformes, Characidae) *Animal Reproduction* .Vol. 3 No. 1 pp. 55-60.

Morisawa, M. and Suzuki, K. (1980). Osmolality and potassium ions: their roles in initiation of sperm motility in teleosts. *Science* Vol. 210 pp.1145-1146.

Olaleye, V.F. (2005). A review of reproduction and gamete management in the African catfish, *Clarias gariepinus* (Burchell 1822). *Ife Journal of Science* Vol. 7 No.1 pp. 63-70.

Omitogun, O. G., Oyeleye, O.O., Betiku, C.O. Ojiokpota, C., Aladele, S.E. and Sarumi, M.B. (2006). Potentials of short-term cryopreserved sperm of the giant African catfish, *Clarias gariepinus*(Burchell, 1822) for aquaculture in Nigeria. In: Olakojo, S.A., Ogunbodede, B.A. and Akande, S.R (Eds) Proceedings of the 31stAnnual Conference of the Genetic Society of Nigeria pp. 141-146. NACGRAB, Moor Plantation, Ibadan, Nigeria. Nov. 6-9, 2006. ISSN 0189-9686.

Omitogun, O.G, Olaniyan, O. F., Oyeleye, O.O., Ojiokpota, C. Aladele, S.E. and Odofin, W T.(2010). Potentials of short term and long term cryopreserved sperm of African giant catfish *(Clarias gariepinus* Burchell 1822) for aquaculture. *African Journal o, Biotechnology* Vol. 9 No. 41. pp. 6973-6982, ISSN 1684-5315. Academic Journals.

Oteme, J., Nunez-Rodriguez, J., Kouassi, C.K. Hom, S. and Agnese, J.F. (1996). Testicular structure, spermatogenesis and sperm cryopreservation in the African clariid catfish, *Heterobranchus longifilis* (Valenciennes, 1840). *Aquaculture Research* Vol. 27 pp. 805-813.

Oyeleye, O. O and Omitogun, O. G. (2007). Evaluation of motility of the short-term cryopreserved sperm of African giant catfish *(Clarias gariepinus). Ife Journal o, Agriculture* Vol. 22 .No.(1) pp.11-16. Ile-Ife, Nigeria. ISSN 0331-6351.

Padhi, B.K. and Mandal, R.K. (1995). Cryopreservation of spermatozoa of two Asian freshwater catfish, *Heteropneutes fossilis* and *Clarias batrachus. Journal of Aquaculture in the Tropics* Vol. 10 pp.23-28.

Riley, K. L. P. (2002). Refrigerated storage and cryopreservation of sperm for the production of Red snapper and Snapper hybrids. *M. Sc Thesis,* Agriculture and Mechanical College, Louisiana University. USA.

SIGMA (1994). Cell counting and cell viability. Saint Louis, MO 63178 USA. Pp. 1634-1635.

Steyn, G.J, Van Vuren, J. H., Schoonbee, H.J. and Chao, N. (1985). Preliminary investigations on the cryopreservation of *Clarias gariepinus* (Clariidae,:Pisces) sperm. *Water SA.* Vol. 11 No.1. pp.15-18.

Steyn, G.J. (1987). The fertilizing capacity of cyopreserved sharptooth catfish, *Clarias gariepinus* sperm. *Aquaculture* Vol. 6 pp. 187-193.

Suquet, M., Dreanno C., Fauvel, C., Cosson, J. and Billard, R. (2000) Cryopreservation of sperm in marine fish. *Aquaculture Research* Vol. 31 pp. 231-243.

Sydenham, D.H.J. (1980). New species of *Clarias* from West Africa. Rev. Zool. *Afr., Vol .XCIV, Fasc.,* Vol. 31 pp. 659-677.

Teugels, G.G. (1986a). Clariidae. In: Daget, J., Gosse, J-P., Thys van den Audenaerde,D.F.E. (editors). Check-list of the Freshwater fishes of Africa, Brussels, MRAC, Tervuren, *ORSTOM,* Paris, pp.66-101.

Teugels, G.G. (1986b). A systematic revision of the African species of the genus Clarias (Pisces:Clariidae). *Annals Musee. Revue. Africain. Contributions.* Vol. 247 pp. 1-199.

Tiersch, T. R., Goudie, C.A. and Carmichael, G.J. (1994). Cryopreservation of channel catfish sperm. Storage in cryoprotectants, fertilization trial and growth of channel catfish produced with cryopreserved sperm. *Transactions of the American Fisheries Society* Vol. 123 pp. 580-586.

Urbanyi B., Horvath, A., Varga, Z. and Horvath, L. (1999). Effect of extenders on sperm cryopreservation of African catfish, *Clarias gariepinus* (Burchell). Aquaculture Research. Vol.30 pp.145-151.

Van der Walt, L.D., Van der Bank, F.H. and Steyn, G.J. (1993). The suitability of using cryopreservation of spermatozoa for the conservation of genetic diversity in African catfish *(Clarias gariepinus). Comparative Biochemistry and Physiology.* Vol. 106A pp.313-318.

Viveiros, A.T., So, N. and Komen, J. (2000). Sperm cryopreservation of African catfish, *Clarias gariepinus;* Cryoprotectants, freezing rates and sperm: egg dilution ratio. *Theriogenology* Vol. 54 pp. 1395-1408.

5

Cryopreservation of Brown Trout (*Salmo trutta macrostigma*) and Ornamental Koi Carp (*Cyprinus carpio*) Sperm

Yusuf Bozkurt[1], İlker Yavas[2] and Fikret Karaca[2]
[1]*Mustafa Kemal University, Faculty of Fisheries, Department of Aquaculture, Hatay*
[2]*Mustafa Kemal University, Faculty of Veterinary Medicine, Department of Reproduction and Artificial Insemination, Hatay*
Turkey

1. Introduction

Cryopreservation is considered as one component in an effective strategy to save endangered species by facilitating the storage of their gametes in gene banks (Gausen, 1993; Akcay et al. 2004). Cryopreservation offers several benefits that in this way stocks can be protected from being totally eliminated due to sudden disease outbreak, natural utilization in hatcheries' production and laboratory experiments can be ensured. Stocks can be maintained more economically and experimental materials for advanced studies, such as gene transfer, can be made more accessible (Chao & Liao, 2001; Tekin et al. 2003).

Cryopreservation techniques involve addition of cryoprotectants, freezing and thawing of sperm samples, all of which may result in some damage to the spermatozoa and may decrease egg fertilization rate (Kopeika et al. 2003). Therefore, before cryopreservation of sperm, a through evaluation of different extender solutions, cryoprotectants, straw sizes and thawing rates is essential to develop optimum cryopreservation protocols for various species (Yavas & Bozkurt, 2011).

The species-specific cryopreservation procedure needs a suitable extender, as undiluted semen is not suitable for long-term preservation. Similarly, addition of optimum amount of cryoprotectant reduces cell damages associated with dehydration, cellular injuries and ice crystal formation (Leung, 1991). Although cryoprotectants help to prevent cryoinjuries during freezing and thawing, they can become toxic to cells when the exposure time and concentration are more (Tekin et al. 2007). In addition, type of cryoprotectants is also very specific to many species. There are no universal extenders and cryoprotectants available that can be used across species.

On the other hand, motility is the most commonly used parameter to evaluate sperm quality in fishes (Billard et al. 1995). This parameter is acceptable so that spermatozoa must be motile to achieve fertilization. Furthermore, sperm motility is an important component of a cryopreservation program in order to prevent poor sperm quality semen samples prior to freezing and to estimate the fertility of the stored sperm after thawing (Akçay et al., 2004;

Bozkurt, 2008). Thawing temperature and duration are also critical factors in the survival of cryopreserved sperm cells (Morris, 1981). Optimal freezig/thawing procedures have not been reported for *Salmo trutta macrostigma* sperm. So, in the present study three different thawing temperatures and thawing durations were also tested related to motility.

For this reason, there is a need to improve techniques on gamete storage and evaluation of sperm quality to facilitate optimization of controlled reproduction in fish (Alavi & Cosson, 2005). Important parameters for cryopreservation include type of extenders and cryoprotectants, dilution ratios, freezing/thawing rates and fertilization rates (Bozkurt et al. 2005).

Salmo trutta macrostigma is a salmonid species occurring in inland water habitats of Southern Europe, Western Asia, Northern Africa and Anatolia (Geldiay & Balik, 1988). It is also critically endangered fish species in inland waters because of illegal fishing, overfishing, and other environmental changes, including hydroelectric plants and pollution. For this reason a biological conservation program has been considered for *Salmo trutta macrostigma* in Turkey. On the other hand, ornamental koi carp is evaluated by its colour and have been used in the selecive propagation programs. These brightly colored koi carps are the result of selective breeding of wild carp. Over centuries a range of pleasing colors, patterns and shapes have been developed for this valuable species. Therefore, reliable methods for brown trout and koi carp sperm cryopreservation could benefit both aquaculture application and conservation of biodiversity.

Therefore, the present study was conducted in order to examine the effect of ionic extenders combined with different cryoprotectants at different ratios and to test the effect of different thawing temperatures and thawing periods on the post-thaw sperm quality of brown trout (*Salmo trutta macrostigma*) and koi carp (*Cyprinus carpio*) and development of a cryopreservation protocol for sperm of this commercially valuable two species.

2. Materials and methods

2.1 Broodstock management

The experiments were carried out spawning season of the brown trout (*Salmo trutta macrostigma*) and koi carp (*Cyprinus carpio*). In the pre-spawning period the mature bown trouts were kept seperately in small ponds under constant environmental conditions. The water temperature ranged 8-10°C during the spawning period. During the experiment, fish were kept under natural photoperiod. Mean water temperature and dissolved oxygen of the broodstock ponds were 8.7±2.46°C and 9.2±7.2 ppm respectively.

The koi carp broodstock was collected from wintering ponds by seining and transported into the hatchery 48 h prior to gamete collection. In the hatchery, male and female broodfish were held seperetely in shadowed tanks (V=1000 L) supplied with continuously (2.5 L min⁻¹) well-aerated water of 24°C. Brown trout and koi carp broodstock were not fed during the experiments.

2.2 Gamete collection

Sperm was collected by gently hand-stripping without anesthesia from mature 10 brown trout males. For koi carp cryopreservation experiments, semen was collected from 5 anesthetized (0.1 g/1 MS 222) males by manual abdominal stripping 12 h after a single injection of 2 mg/kg of carp pituitary extract (CPE) at 20-22 °C water temperature. Eggs were collected by hand stripping 10-12 h after a double injection of 3.5 mg/kg of CPE. The first injection, 10% (0.35 mg/kg) CPE was given 10 h before the second (3.15 mg/kg).

For sperm collection, the urogenital papilla's of mature male fishes were carefully dried and sperm was hand-stripped directly into test tubes. Following sperm collection, the tubes containing sperm were placed in a styrofoambox containing crushed ice (4°C). Contamination of sperm with water, urine or faeces was carefully avoided. Sperm was transported to the laboratory within 15 min. For collection of eggs from koi carps, females were wiped dry, stripped by gentle abdominal massage and the eggs from each female were collected in a dry metal bowl. Eggs were checked visually and only those lots of homogenous shape, colour and size were used in the fertilization experiments.

2.3 Determination of fresh sperm quality parameters

Motility was estimated subjectively using light microscope (Olympus, Japan) with a x400 magnification. Samples were activated by mixing 1 μl of sperm with 20 μl activation solution (0.3% NaCl) on a glass slide. The percentage of motility was defined as the percentage of spermatozoa moving in a forward motion every 20% motile increment (i.e., 0, 20%, 40%, 60%, 80%, and 100%) (Vuthiphandchai & Zohar, 1999). Motility measurements were performed within 15 s. after activation. Sperm cells that vibrated in place were not considered to be motile. Sperm motility was estimated with three replicates of samples. For cryopreservation experiments, samples below 80% motile spermatozoa were discarded. Duration of sperm motility was determined using a sensitive chronometer (sensitivity: 1/100 s) by recording the time following addition of the activation solution to the sperm samples.

Spermatozoa density was determined according to the haemacytometric method. Sperm was diluted at ratio of 1:1000 with Hayem solution (5g Na_2SO4, 1g NaCl, 0.5g $HgCl_2$, 200 mL bicine) and density was determined using a 100 μm deep Thoma haemocytometer (TH-100, Hecht-Assistent, Sondheim, Germany) at 400x magnification with Olympus BX50 phase contrast microscope (Olympus, Japan) and expressed as spermatozoa $x10^9$ mL^{-1} (three replicates). Counting chambers were always kept in a moist atmosphere for at least 10 min before cell counting. Sperm pH was measured using indicator papers (Merck, 5.5-9) within 30 min of sampling.

2.4 Experiment 1 - Brown trout (*Salmo trutta macrostigma*)

Collected sperm from 10 males that showing >80 motility was pooled into equal aliquots according to the required semen volume and sperm density to eliminate effects of individual variability of gamete donors. Semen and extenders were kept at 4°C prior to dilution. Pooled semen was diluted at 1:3 ratio (semen/extender) with extender containing 4.68 g l- NaCl, 2.98 g l- KCl, 0.11 g l- CaCl₂ and Trizma-HCl 3.15 g l- in distilled water; pH 9.0 (Billard & Cosson, 1992). The extender contained methanol and egg yolk at ratios of 5%, 10% and 15% separately. Dilution of semen with extender resulted in sperm concentrations of around 2.5x10⁹ cells/ml extender that was enough to avoid damage due to sperm compression during freezing and thawing (Lahnsteiner, 2000). Following sperm suspension was equilibrated for 10 min at 4°C.

Within 1 h after sperm collection, the diluted semen samples were drawn into 0.25mL plastic straws (IMV, France). The open end of straws were sealed with polyvinyl alcohol (PVA). Following, the straws were placed on a styrofoam rack that floating on the surface of liquid nitrogen in a styrofoam box. The straws were frozen in liquid nitrogen vapour 4 cm above of the liquid nitrogen surface (temperature of styroframe surface was about -140°C) for 10 min. Following, the straws were plunged into the liquid nitrogen (-196°C) and stored for several days. For thawing, straws were thawed at 30°C for 10 s by gentle agitation in water bath. Thawed sperm was activated using pond water.

On the other hand, post-thaw sperm quality tests were carred out to evaluate motility rate and duration of motility. For this aim, sperm motility rate and duration of motility values following cryopreservation in the same ionic extender containing 15% egg yolk were determined. Sperm was thawed at 25°C, 35°C or 45°C for 5s, 15s or 25s and activated in 0.3% NaCl and 1% NaHCO₃.

2.5 Experiment 2 - Ornamental koi carp (*Cyprinus carpio*)

Collected semen from the 5 males that showing >80 motility was pooled into equal aliquots according to the required semen volume and sperm density needed to eliminate effects of individual variability of the donors. Semen and extenders were kept at 4°C, then diluted at a ratio of 1:3 (semen/extender) with 3 different extenders containing 10% DMSO. Extender 1 contained 5.8 g/L NaCl, 0.2 g/L KCl, 0.22 g/L CaCl₂, 0.04 g/L MgCl₂6H₂O, 2.1 g/L NaHCO₃, 0.04 g/L NaH₂PO₄.2H₂O, 3.75 g/L glycine (Ravinder, et al. 1997). Extender 2 contained 300 mM glucose and 10% egg yolk pH:8 (Tekin et al. 2003) and extender 3 contained 4.68 g/L NaCl, 2.98 g/L KCl, 0.11 g/L CaCl₂, 3.15 g/L Tris-HCl, pH:9 (Billard & Cosson, 1992).

The diluted samples were drawn into 0.25 ml plastic straws (IMV, France) and were sealed with polyvinyl alcohol (PVA). Having been diluted, the samples were equilibrated for 10 min at 4°C. After equilibration, the straws were placed on a styrofoam rack that floated on the surface of liquid nitrogen in a styrofoam box. The straws were frozen in liquid nitrogen vapour 3 cm above the surface of liquid nitrogen (-140°C) for 10 min. After 10 min the straws were plunged into the liquid nitrogen (-196°C) and stored for several days. For thawing, the straws were removed from liquid nitrogen and immersed in 30°C water for 10 seconds. Thawed sperm was activated using 0.3% NaCl and observed under microscope for determination of spermatozoa motility and motility durations.

For fertilization experiments, pooled eggs from 3 mature females were used to determine fertilization rates. Egg samples (about 100 eggs) were inseminated in dry Petri dishes with fresh sperm or frozen sperm immediately after thawing at a spermatozoa:egg ratio of 1×10^5: 1. Eggs were inseminated by the dry fertilization technique using a solution of 3 g urea and 4 g NaCl in 1 L distilled water. The sperm and eggs were slightly stirred for 30 min, washed with hatchery water (24°C; 9 mg/l O₂), and gently transferred to labeled Zuger glass incubators with running water (24°C) where they were kept until hatching (3-4 d). Living and dead eggs were counted in each incubator during incubation and dead eggs were removed. When the fertilized eggs developed to embryos at the gastrula stage, the fertilization rate (number of gastrula stage embryos/number of total eggs) was calculated.

2.6 Statistical analysis

Results are presented as means±SE. Differences between parameters were analyzed by repeated analysis of variance (ANOVA). Significant means were subjected to a multiple comparison test (Duncan) for post-hoc comparisons at a level of α=0.05. All analyses were carried out using SPSS 10 for Windows statistical software package.

3. Results

3.1 Fresh sperm quality parameters

In brown trout fresh semen volumes were rather variable and ranged from 9 to 17 ml and mean volume was 12.6±4.28 mL. Motility values were ranged from 75% to 90%. Samples that motility values were below than 80% were not used for the cryopreservation experiments. The mean motility value of fresh sperm samples were 84.5±7.59%. Mean spermatozoa movement duration (s), sperm density x10^9/mL and pH values were achieved as 57.4±3.8 s, 24.8±4.62 x10^9/mL and 7.28±2.46 respectively.

In koi carp mean fresh semen volume, spermatoza motility, motility duration, spermatozoa density and pH values of the collected fresh milt samples were determined as 6.2±4.7 ml, 85.4±2.4%, 125.2±3.5 s, 22.8 x 10^9 mL^{-1} and 7.4±3.7, respectively.

3.2. Experiment 1 - Brown trout (*Salmo trutta macrostigma*)

Post-thaw motility of sperm cryopreserved in ionic extender containing two different cryoprotectants at three different ratios is shown in Table 1. There were significant effect of cryoprotectants on motility rates. Sperm samples cryopreserved in the extenders containing egg yolk yielded greater post-thaw motility rates than methanol containing extenders. Sperm frozen with extender containing 15% egg yolk had the highest post-thaw motility. Differences between the post-thaw motility values were significant (P<0.05).

Methanol (5%)	Methanol (10%)	Methanol (15%)	Egg Yolk (5%)	Egg Yolk (10%)	Egg Yolk (15%)
10.6±4.57Af	15.2±5.80Ae	17.4±4.72Ae	40.5±3.27Aab	42.3±6.1Aa	45.3±4.27Aa
7.5±2.69Ae	9.6±3.37Be	12.3±5.24Ad	30.6±2.86Bb	35.4±4.17Abab	40.2±5.36Aba
5.4±2.73Ae	7.2±3.79Be	10.5±4.27Abd	25.7±4.69BCbc	30.2±5.29Bb	37.8±8.29Ba

Means followed by different superscripts (lowercase for lines and uppercase for columns within the same sperm feature) are different (p<0.05). (mean±SE, n=3).

Table 1. Post-thaw motility (%) of brown trout sperm cryopreserved with different cryoprotectants.

It was observed that a decrease in motility duration occurred following cryopreservation. The longest post-thaw motility longevity was also achieved with extender containing 15%

egg yolk as 54.2±3.46 s. Differences between the means of motility durations were significant (P<0.05). (Table 2).

Methanol (5%)	Methanol (10%)	Methanol (15%)	Egg Yolk (5%)	Egg Yolk (10%)	Egg Yolk (15%)
20.3±2.57Ad	24.5±2.47Acd	28.6±3.46Ac	40.2±1.29Aa	42.6±4.57Aa	46.4±2.38Ba
15.23±4.39Abf	19.6±4.39Abef	23.4±3.47Abe	37.2±3.45Ac	45.3±5.39Ab	54.2±3.46Aa
12.3±4.17Bd	15.7±1.28Bd	20.4±8.25Bcd	20.4±4.59Bcd	25.3±2.48Bc	32.5±5.27Cb

Means followed by different superscripts (lowercase for lines and uppercase for columns within the same sperm feature) are different (p<0.05). (mean±SE, n=3).

Table 2. Post-thaw longevity (s) of brown trout sperm cryopreserved with different cryoprotectants.

Sperm motility rate (Figure 1) and longevity of motility (Figure 2) values following cryopreservation in the ionic extender containing 15% egg yolk were determined. Sperm was thawed at 25°C, 35°C or 45°C for 5s, 15s or 25s and activated in 0.3% NaCl and 1% NaHCO$_3$.

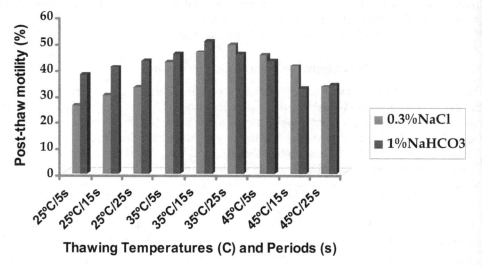

Fig. 1. Post-thaw motility (%) of brown trout sperm thawed at different degrees, periods and activating agents.

Post-thaw sperm motility rates were affected by thawing rates and activation agents and ranged from 25% to 50%. Also, the activating agents affected the duration of motility. All sperm samples triggered in 1% NaHCO$_3$ were motile for a longer period (32-57 s) compared with samples triggered in 0.3% NaCl (24-53 s). Differences between the post-thaw motility and longevity values were significant (P<0.05).

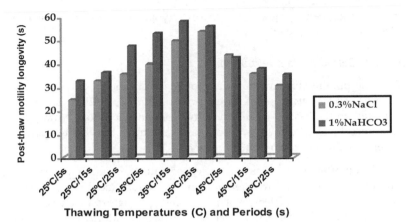

Thawing Temperatures (C) and Periods (s)

Fig. 2. Post-thaw motility longevity (s) of brown trout sperm thawed at different degrees, periods and activating agents.

3.3 Experiment 2 - Ornamental koi carp (*Cyprinus carpio*)

Effect of three different extenders containing 10% DMSO on the post-thaw motility and movement duration, fertilization and hatching rates of koi carp are shown in Table 3. Mean post-thaw motility of koi carp sperm was 75.3±6.4% while the best motility was determined as 85%. The overall mean fertilization rate was determined as 99.2±0.72 while the best fertilization rate was determined as 100%. The highest hatching rate was determined as 50% in all experimental groups. Motility features and hatching rates of cryopreserved koi carp sperm was statistically different between the experimental groups (p<0.05).

Extenders	Post-thaw motility (%)	Post-thaw motlity duration (s)	Fertilization rates (%)	Hatching rates (%)
E1	75.2±0.4[a]	27.5±1.2[a]	99.6±0.5	42.5±1.9[b]
E2	78.6±0.7[b]	32.9±0.4[b]	99.7±0.5	46.2±0.7[c]
E3	72.3±0.2[a]	25.2±0.6[a]	98.3±1.2	37.4±0.2[a]
Control	-	-	99.8±0.2	86.2±0.4[d]

Means followed by different superscripts are different (p<0.05). (mean±SE, n=3).

Table 3. Effect of different extenders on post-thaw motility, fertilization and hatching rates of koi carp sperm.

4. Discussion

Successful cryopreservation of fish spermatozoa depends on a range of factors including the collection of high quality sperm, equilibration conditions, choice of cryoprotectant medium, cooling/thawing regimes, and conditions for fertilization. Even though some general rules can be applied to any fish species, optimization of the protocol is needed for each individual species (Kopeika et al. 2007). Several factors have affected post-thaw quality of cryopreserved sperm from both brown trout (*Salmo trutta macrostigma*) and ornamental koi carp (*Cyprinus carpio*). The results obtained in the present study contribute significantly improve the development protocol of sperm cryopreservation in brown trout and ornamental koi carp at large scale.

4.1 Brown trout (*Salmo trutta macrostigma*)

The results of the present study demonstrate for the first time cryopreservation of brown trout (*Salmo trutta macrostigma*) sperm. In the present study, post-thaw sperm quality was initially evaluated on the basis of sperm motility score and duration of motility for brown trout. For this aim, the effect of two cryoprotectants and three thawing temperatures on the post-thaw sperm quality of brown trout were assessed.

Motility is induced after the spermatozoa released into the aquatic environment during natural reproduction or after transfer to an activation medium during controlled reproduction (Alavi & Cosson, 2006). When salmonid spermatozoa are released into water they have a brief period of sperm activity between 20 and 40s (Morisawa & Morisawa, 1986). A better knowledge of the characteristics of fresh sperm motility is necessary to evaluate sperm quality in commercial hatcheries before artificial reproduction and in laboratories before experiments. Preliminary examination of fresh sperm was carried out in order to determine the relationship between sperm motility and seminal plazma composition of *Salmo trutta macrostigma* sperm (Bozkurt et al. 2011a).

Comparison of different cryoprotectant recipes and freeze-thaw protocols are difficult when each treatment tested for the ability of sperm to fertilise eggs. Cryoprotectants can suppress most cryoinjuries when used higher concentrations but at the same time it can become toxic to the cells. Therefore, a suitable concentration was needed for the development of a cryopreservation protocol. Methanol has been used successfully for sperm cryopreservation in African catfish (*Clarias gariepinus*) (Burchell) (Steyn & Van Vuren, 1987), tilapia (*Sarotherodon mossambicus*) (Peters) (Rana & McAndrew, 1989), bagrid catfish (*Mystus nemurus*) (Muchlisin et al., 2004) and salmonid fish (Lahnsteiner et al., 1996). Mansour et al. (2006) showed that 10% methanol was more effective as a cryoprotectant for Arctic char spermatozoa than 10% DMA or 10% DMSO when used with a glucose diluent. However, the effects of higher levels of methanol cryoprotectant were not investigated. In the present study, brown trout semen in an extender containing 15% egg yolk resulted in the highest overall percentage of sperm motility.

In addition, penetrating cryoprotectants could affect the percentage of motile sperm. In salmonids, some authors reported that higher post-thaw motility from methanol than from DMSO and other cryoprotectants (Mansour et al. 2006). In the present study, methanol and egg yolk have statistically significant effect on the percentage of sperm motility. On the other hand, it should be noted that egg yolk achieved better results than methanol for cryopreservation of brown trout sperm. Extender containing glucose, egg yolk and DMSO described by Alderson and McNeil (1984) gave good results in cryopreservation experiments with large straws. Baynes and Scott (1987) also reported that egg yolk is a valuable component in extenders for salmonid sperm cryopreservation. Furthermore, the addition of egg yolk to the medium interferes with the good visualization of spermatozoa during the motility rate analysis. With this in mind, we have tested several extender/cryoprotectant combinations with the addition of egg yolk that preserve sperm during storage and yet allow good visualization during motility analysis.

On the other hand, thawing temperature also play an important role in the post-freeze semen quality of fish (Wayman et al., 1998). Generally, thawing rates should be high to avoid recrystallization (Lahnsteiner, 2000). Significant post-thaw motility was determined when brown trout sperm was thawed at temperature of 35°C in the present study. According to the results of the present study, it was shown that higher temperatures are necessary to recover membrane stability or metabolism of spermatozoa. Also it appears that either recrystallization and ice crystal formation during thawing were reduced or avoided by this thawing procedure, or enzymatic activities were the best reactivated (Lahnsteiner, 2000). Although thawing from -196°C to 4°C is generally considered as critical phase because of potential recrystallization, the process was similar for all species. Furthermore, the two activating agents (0.3% NaCl and 1% NaHCO$_3$) tested did not affect post-thaw motility rates or quality motility score, although, in general, higher scores were observed when 1% NaHCO$_3$ was used in the present study. Duration of motility was significantly higher when 1% NaHCO$_3$ was used as an activating agent.

4.2 Ornamental koi carp (*Cyprinus carpio*)

The main purpose of the current experiment was to develop an appropriate protocol for ornamental koi carp (*Cyprinus carpio*) sperm cryopreservation to increase sperm availability outside the breeding season. By banking male gametes when they are abundant, most efforts can be devoted to raising healthy female broodstock and obtaining good quality eggs within a short captivity culture period. Through cryopreservation, a sperm repository can also be established for all males in captivity or from the wild. Such repository is important to maintain the genetic diversity to avoid inbreeding or loss of heterozygosity for captive breeding programs as well as possible future stock enhancement in the wild (Cabrita et al., 2009).

In the present experiment, koi carp males gave sperm characterized by good spermatozoa density and percent of motility. Such sperm should be used for cryopreservation experiments when considering minimization of artificial selection and sperm competition during hatchery operations in order to maintain the greatest biodiversity (Campton, 2004).

During the cryopreservation process, one of the important issue is the use of cryoprotectants, which role is to prevent cell damage during the freezing and thawing steps. Several cryoprotectants have been used for fish sperm cryopreservation, including methanol, ethylene glycol and dimethyl sulphoxide (DMSO); however, DMSO is reported to be the most efficient to cryopreserve fish spermatozoa (Anel & Cabrita, 2000) due mainly to its small molecular size, which allows it to enter and exit the spermatic cell easily (Tiersch et al. 1998).

The best fertilization rate obtained with extender II with 99.7% eyeing rate in koi carp. These results can be explained by the presence of 10% DMSO as a cryoprotectant in this extender. It can be concluded that DMSO has higher permeability by permeating into cell, causing reduced ice crystal formation for koi carp sperm. On the other hand, Lahnsteiner et al., (1996) used 10% methanol, 10% DMSO, 10% DMA, 5% glycerol and mixture of 5% DMSO and 1% glycerol for semen cryopreservation of the grayling (*Thymallus thymallus*) and the Danube salmon (*Hucho hucho*), which methanol showed the highest fertilization rates in relation to control 95.3% and 91.1% for grayling and Danube salmon, respectively.

Cryopreservation protocol carried out in the present study with a 1:100,000 egg: spermatozoa ratio, almost the same fertilization efficiency was obtained whether frozen or fresh semen was used. This may be due to differences in extender, cryoprotectant, equilibration, egg quality, or protocol. In the present study, the interaction between the percentage of motile post-thaw sperm and fertilizing capacity was highly positive, similar to results in common carp (Linhart et al., 2000), African catfish (Rurangwa et al., 2001) and grass carp (Bozkurt et al., 2011b).

On the other hand, a wide range of temperatures used to thawed cryopreserved sperm with temperatures from refrigeration (4°C) to 80°C were reported (Lahnsteiner et al., 2000). A fast thawing temperature decreases the recrystallization effect in the spermatic cells and therefore diminishes the membrane damage (Tiersch et al.1998). Higher temperature such as 30°C were also used to thaw cyprinid semen in several studies (Stoss & Hotz, 1983) that similar with the pesent study.

5. Conclusion

It can be concluded that the cryopreservation protocol developed in this study is rather effective and brown trout (*Salmo trutta macrostigma*) and ornamental koi carp (*Cyprinus carpio*) sperm can be successfully cryopreserved. It seems that cryopreservation of brown trout sperm with ionic extenders containing 15% egg yolk is rather effetive on post-thaw sperm quality. In addition, based on the results obtained from this study, it is possible to suggest that sperm cryopreserved with ionic extender containing 10% DMSO packed in 0.25 mL volume straws and thawed at 30°C are the most suitable conditions to retain the sperm quality in koi carp having optimal sperm motility, duration of motility as well as high fertility percentages close to the values obtained with fresh sperm.

This study can help establish a frozen sperm bank for the conservation of genetic material of the brown trout and koi carp. On the other hand, additional research is needed on the effects of cryoprotectants, protective agents and freezing technique in cryopreservation on malformations, survival and condition of progeny produced with cryopreserved spermatozoa of brown trout and koi carp.

6. Acknowledgements

This research was financed by Research Fund of Musafa Kemal University (1005-M-0111). The authors would like to thank the staff of the Fish Production Station of General Directorate of National Parks in Tarsus and State Hydraulic Works (SHW) Fish Production Station in Adana, Turkey.

7. References

Akcay, E.; Bozkurt, Y.; Secer, S. & Tekin, N. (2004). Cryopreservation of mirror carp semen. *Turkish Journal of Veterinary and Animal Sciences*, 28 (5), 837-843.

Alavi, S.M.H. & Cosson, J. (2005): Sperm motility in fishes: (I) Effects of temperature and pH: a review. *Cell Biol. Intern.* 29, 101–110.

Alavi, S.M.H. & Cosson, J. (2006). Sperm motility in fishes: II. Effects of ions and osmolality. *Cell Biol. Int.*, 30, 1-14.

Alderson, R. & Macneil, A.J. (1984). Preliminary investigation of cryopreservation of Atlantic salmon *Salmo salar* and its application to commercial farming. *Aquaculture*, 43: 351–354.

Anel, I. & Cabrita, E. (2000). Effect of external cryoprotectants as membrane stabilizers on cryopreserved rainbow trout sperm. *Theriogenology*, 13: 623-635.

Baynes, S.M. & Scott, A.P. (1987). Cryopreservation of rainbow trout spermatozoa: the influence of sperm quality, egg quality and extender composition on post-thaw fertility. *Aquaculture*, 66:53-67.

Billard, R. & Cosson, M.R. (1992). Some problems related to the assessment of sperm motility in freshwater fish. *Journal of Experimental Zoology*, (261) 122-131.

Billard, R.; Cosson, J.; Crim, L.W. & Suquet, M. (1995). Sperm Physiology and Quality. In: *Broodstock Management and Egg and Larval Quality*. Blackwell Science, Oxford, pp: 25-52.

Bozkurt, Y.; Akçay, E.; Tekin, N. & Secer, S. (2005). Effect of freezing techniques, extenders and cryoprotectants on the fertilization rate of frozen rainbow trout *(Oncorhynchus mykiss)* sperm. *The Israeli Journal of Aquaculture-Bamidgeh*, 57 (2), 125-130.

Bozkurt, Y. (2008). Physical and biochemical properties of *Salmo trutta abanticus* semen. *Indian Veterinary Journal*, 85 (3), 282-284.

Bozkurt, Y; Ogretmen, F.; Kokcu, O. & Ercin, U. (2011a). Relationship between seminal plasma composition and sperm quality parameters of the *Salmo trutta macrostigma* (Dumeril, 1858) semen: with emphasis on sperm motility. *Czech Journal of Animal Science*, 56 (8), 355-364.

Bozkurt, Y.; Yavas, I.; Ogretmen, F.; Sivaslıgil, B.; & Karaca, F. (2011b). Effect of glycerol on fertility of cryopreserved grass carp (*Ctenopharyngodon idella*) sperm. *The Israeli Journal of Aquaculture-Bamidgeh*, IIC:63.2011.635.

Cabrita, E.; Engrola, S.; Conceicao, L.; Pousaoferreira, P. & Dinis, M.T. (2009). Successful cryopreservation of sperm from sex-reversed dusky grouper, Epinephelus marginatus. *Aquaculture*, 287: (1-2), 152-157.

Campton, D. F. (2004). Sperm competition in salmon hatcheries: the need to institutionalize genetically benign spawning protocols. *Transactions of American Fisheries Society*, 133, 1277–1289.

Chao, N.H. & Liao, I.C. (2001). Cryopreservation of finfish and shellfish gametes and embryos. *Aquaculture*, 197:161-189.

Gausen, D. (1993). The Norwegian gene bank programme for Atlantic salmon (*Salmo salar*). *In*: J.C. Cloud and G.H. Thorgaad (eds.), *Genetic conservation of salmonid fishes*. *Plenum*. New York, USA. pp.181-187.

Geldiay, R. & Balik, S. (1988): *Turkiye tatlısu balıkları* (Freshwater fishes in Turkey). Ege Univ. Fen Fak. Kitaplar Serisi. 97, pp. 519.

Kopeika, J.; Kopeika, E.; Zhang, T. & Rawson, D.M. (2003). Studies on the toxicity of dimethyl sulfoxide, ethylene glycol, methanol and glycerol to loach (Misgurnus fossilis) sperm and the effect on subsequent embryo development Cryo Letters. 24 (6) : 365-374.

Kopeika, E.; Kopeika, J. & Zhang, T. (2007). Cryopreservation of fish sperm . Methods Mol Biol. 368 : 203 - 217.

Lahnsteiner, F.; Weismann, T. & Patzner, R.A. (1996). Cryopreservation of semen of the grayling (*Thymallus thymallus*) and the Danube salmon (*Hucho hucho*). *Aquaculture*, 144:265-274.

Lahnsteiner, F. (2000). Semen cryopreservation in the salmonidae and in the northern pike. *Aquaculture Research*, 31:245-258.

Lahnsteiner, F.; Berger, B.; Horvath, A.; Urbanyi, B. & Weismann T. (2000). Cryopreservation of spermatozoa in cyprinid fishes. *Theriogenology*, 54: 1477–1498.

Leung, L.K.P. (1991). Principles of biological cryopreservation. In. *Fish evolution and systematics: Evidence from spermatozoa.* Jamieson, B.M.G. (Ed). Cambridge University Press, Cambridge. 231-244.

Linhart O.; Rodina M. & Cosson, J. (2000). Cryopreservation of sperm in common carp sperm motility and hatching success of embryos. *Cryobiology,* 41:241-250.

Mansour, N.; Richardson, G.F. & McNiven, M.A. (2006). Effect of extender composition and freezing rate on post-thaw motility and fertility of Arctic char, *Salvelinus alpinus* (L.), spermatozoa. *Aquaculture Research,* 37: 862– 868.

Morisawa, S. & Morisawa, M. (1986). Acquisition of potential for sperm motility in rainbow trout and chum samon. *The Journal of Experimental Biology,* 126: 89-96.

Morris, G.J. (1981). *Cryopreservation: an introduction to cryopreservation in culture collections.* Cambridge (England): Institute of Terrestrial Ecology.

Muchlisin, Z.A.; Hashim, R. & Chong, A.S.C. (2004). Preliminary study on the cryopreservation of tropical bagrid catfish (*Mystus nemurus*) spermatozoa; the effect of extender and cryoprotectant on the motility after short-term storage. *Theriogenology,* 62, 25–34.

Rana, K.J. & McAndrew, B.J. (1989). The viability of cryopreserved tilapia spermatozoa, *Aquaculture,* 76: 335-345.

Ravinder, K.; Nasaruddin, K.; Majumdar, K. C. & Shivaji, S. (1997). Computerized analysis of motility, motility patterns and motility parameters of carp following short-term storage of semen. *Journal of Fish Biology,* 50: 1309–1328.

Rurangwa, E.; Volckaert, F.A.M.; Huyskens, G.; Kime, D.E. & Ollevier, F. (2001). Quality control of refrigerated and cryopreserved semen using computer-assisted sperm analysis (CASA), viable staining and standardized fertilisation in African catfish (*Clarias gariepinus*). *Theriogenology,* 55:751-769.

Secer, S.; Tekin, N.; Bozkurt, Y.; Bukan, N. & Akçay, E. (2004). Correlation between biochemical and spermatological parameters in rainbow trout (*Oncorhynchus mykiss*) semen. *The Israeli Journal of Aquaculture-Bamidgeh,* 56 (4), 274-280.

Steyn, G.J. & Van Vuren, J.H.J. (1987). The fertilising capacity of cryopreserved sharptooth cattish (*Clarias gariepinus*) sperm. *Aquaculture,* 63: 187-193.

Stoss, J. & Holtz, W. (1983). Successful storage of chilled rainbow trout (*Salmo gairdneri*) spermatozoa for up to 34 days, *Aquaculture,* 31: 269-274.

Tekin, N.; Secer, S.; Akcay, E. & Bozkurt, Y. (2003). Cryopreservation of rainbow trout (*Oncorhynchus mykiss*) semen. *The Israeli Journal of Aquaculture-Bamidgeh,* 55 (3), 208-212.

Tekin, N.; Secer, S; Akcay, E.; Bozkurt, Y. & Kayam, S. (2007). Effects of glycerol additions on post-thaw fertility of frozen rainbow trout sperm, with an emphasis on interaction between extender and cryoprotectant. *Journal of Applied Ichthyology,* 23 (1): 60-63.

Tiersch, T.R.; Williamson, J.H.; Carmichael, G.J. & Gorman, O.T. (1998). Cryopreservation of sperm of the endangered razorback sucker. *Transactions of the American Fisheries Society,* 127: 95-104.

Vuthiphandchai, V. & Zohar, Y. (1999). Age-related sperm quality of captive striped bass, *Morone saxatilis. Journal of World Aquaculture Society,* 30: 65–72.

Wayman, W.R.; Tiersch, T.R. & Thomas, W.R. (1998). Refrigerated storage and cryopreservation of sperm of red drum, *Sciaenops ocellatus* L. *Aquaculture Research,* 29: 267-273.

Yavas, I. & Bozkurt, Y. (2011). Effect of different thawing rates on motility and fertilizing capacity of cryopreserved grass carp (*Ctenopharyngodon idella*) sperm. *Biotechnology and Biotechnological Equipment,* 25 (1), 2254-2257.

Part 2

Cryopreservation of Plants

6

Plant Cryopreservation

R.K. Radha, William S. Decruse and P.N. Krishnan

Plant Biotechnology and Bioinformatics Division
Tropical Botanic Garden and Research Institute
Palode, Thiruvananthapuram, Kerala,
India

1. Introduction

Two basic approaches to conservation of plant genetic resources are *ex situ* and *in situ* conservation. *Ex situ* conservation includes seed storage, *in vitro* storage, DNA storage, pollen storage, field genebanks and botanical gardens while the *in situ* approach encompasses genetic reserves, on farm and home garden conservation.

Cryopreservation is a part of biotechnology. Biotechnology plays an important role in international plant conservation programs and in preservation of the world's genetic resources (Bajaj, 1995; Benson, 1999). Advances in biotechnology provide new methods for plant genetic resources and evaluation (Paunesca, 2009). Cryopreservation, developed during the last 25 years, is an important and the most valuable method for long-term conservation of biological materials. The main advantages in cryopreservation are simplicity and the applicability to a wide range of genotypes (Engelmann, 2004). This can be achieved using different procedures, such as pre-growth, desiccation, pregrowth-desiccation, vitrification, encapsulation-vitrification and droplet-freezing (Engelmann, 2004). Cryopreservation involves storage of plant material (such as seed, shoot tip, zygotic and somatic embryos and pollen) at ultra-low temperatures in LN (-196°C) or its vapor phase (-150°C). To avoid the genetic alterations that may occur in long tissue cultures storage, cryopreservation has been developed (Martin *et al.*, 1998). At this temperature, cell division, metabolic, and biochemical activities remain suspended and the material can be stored without changes and deterioration for long time. Walters *et al.* (2009) proposed that this assumption, based on extrapolations of temperature-reaction kinetic relationships, is not completely supported by accumulating evidence that dried seeds can deteriorate during cryogenic storage. After 30 years of cryogenic storage, seeds of some species exhibited quantitatively lower viability and vigor. In cryopreservation method, subcultures are not required and somaclonal variation is reduced. Advantages of cryopreservation are that germplasm can be kept for theoretically indefinite time with low costs and little space. Besides its use for the conservation of genetic resources, cryopreservation can also be applied for the safe storage of plant tissues with specific characteristics. Different types of plant cell, tissues and organs can be cryopreserved. Cryopreservation is the most suitable long-term storage method for genetic resources of vegetatively maintained crops (Kaczmarczyk *et al.*, 2008). For vegetatively propagated species, the best organs are shoot apices excised from *in vitro* plants. Shoot apices or meristems cultures are suitable because of virus-free plant production, clonal propagation, improving health status, easier recovery and less mutation (Scowcroft, 1984). Seed and field collections have been the only proper for the long-term germplasm conservation of woody species, while a large number of forest angiosperms have recalcitrant seeds with a very limited period of conservability. The species, which are mainly vegetatively propagated, require the conservation of huge number of accessions (Panis and Lambardi, 2005). The storage of this huge number needs large areas of land and high running costs. Preservation of plant germplasm is part of any plant breeding program. The most efficient and economical way of germplasm storage is the form of seeds. However, this kind of storage is not always feasible because 1) some seeds deteriorate due to invasion of pathogens and insects, 2) some plants do not produce seeds and they are propagated vegetatively, 3) some seeds are very heterozygous thus, not proper for maintaining true-to-type genotype, 4) seeds remain viable for a limited time, and 5) clonally propagated crops such as fruit, nut, and many root and tuber vegetables cannot be stored as seed (Chang and Reed, 2001; Bekheet et al., 2007). Cryopreservation offers a good method for conservation of the species, especially woody plant germplasm (Panis and Lambardi, 2005). Cryostorage of seeds in LN was initially developed for the conservation of genetic resources of agriculturally important species (Rajasekharan, 2006). The development of simple cryostorage protocols for orthodox seeds has allowed cryopreservation of a large number of species at low cost, significantly reducing seed deterioration in storage (Stanwood, 1987). Only few reports are available on the application of cryopreservation on seeds of wild and endangered species and medicinal plants (Rajasekharan, 2006). New cryobiological studies of plant materials has made cryopreservation a realistic tool for long-term storage, for tropical species, which are not intrinsically tolerant to low temperature and desiccation, has been less extensively investigated (Rajasekharan, 2006). Cryopreservation has been applied to more than 80 plant species (Zhao et al., 2005). Number of species, which can be cryopreserved has rapidly increased over the last several years because of the new techniques and progress of cryopreservation research (Rajasekharan, 2006). The vitrification/one-step freezing and encapsulation dehydration methods have been applied to an increasing number of species (Panis and Lambardi, 2005). A new method, named encapsulation- vitrification is noteworthy (Sakai, 2000). These techniques have produced high levels of post-thaw and minor modifications (Rajasekharan, 2006). In cryopreservation, information recording such as type and size of explants, pretreatment and the correct type and concentration of cryoprotectants, explants water content, cryopreservation method, rate of freezing and thawing, thawing method, recovery medium and incubation conditions is very important (Reed, 2001 ; González-Benito et al., 2004; Bekheet et al., 2007). All germplasm requires safe storage because even exotic germplasm without obvious economic merit may contain genes or alleles that may be needed as new disease, insect, environmental, or crop production problems arise (Westwood, 1989). It is important to record also the recovery percentage after a short conservation period. A major concern is the genetic stability of conserved material.

For many plant species which produce orthodox seeds, i.e. which can be dehydrated extensively and stored dry at low temperature, the emphasis for genetic resource conservation will be on seed/embryo storage. Recalcitrant seeds cannot tolerate desiccation to moisture content that would permit exposure to low temperature. They are often large with considerable quantities of fleshy endosperm. Therefore, recent investigations have identified species displaying an intermediate form of seed/embryo storage. As regards the balance of techniques employed within complementary strategies developed for conserving the genetic resources of these problems species, the emphasis in the case of non-orthodox (intermediate/ recalcitrant) forest tree species will be on *in situ* conservation in genetic reserves, while for species which are propagated vegetatively the emphasis will be on *ex situ* conservation techniques, including field genebank and *in vitro* storage. However it is essential to recognize that owing to various problems and limitations encountered with both genetic reserves and field genebanks, cryopreservation currently offers the only safe and cost effective option for the long-term conservation of genetic resources of these problem species. Significant progress has been made during the past 10 years in the area of plant cryopreservation with the development of various efficient cryopreservation protocols. An important advantage of these new techniques is their operational simplicity, since they will be applied mainly in developing tropical countries where the largest part of genetic resources of problem species is located. Encouraging results in medicinal plants have been published in recent years which present extensive list of plant species whose embryos and or embryonic axes have been successfully cryopreserved (Kartha and Engelmann 1994, Pence 1995, Engelmann *et al* 1995).

In comparison with results obtained with vegetatively propagated species, it is clear that research is still at a very preliminary stage for recalcitrant seeds. The desiccation technique is mainly employed for freezing embryos and embryonic axes, the survival achieved are extremely uneven. And also survival is often limited and regeneration often restricted to callusing or incomplete development of plantlets. In only a limited number of cases, the whole plants have been regenerated from cryopreserved material (Chin and Pritchard 1988, Assy Bah and Engelmann 1992). Seeds and embryos of recalcitrant species also display various characteristics which make their cryopreservation difficult. One of the characteristics of recalcitrant seeds is that there is no arrest in their development, as with orthodox seeds. It is very difficult to select seeds at a precise developmental stage, even though this parameter is often of critical importance to achieve successful cryopreservation. Seeds of many species are too large to be frozen directly and embryos or embryonic axes have to be employed. However, embryos are often very complex tissue composition which display differential sensitivity to desiccation and freezing, the root pole seeming more resistant than the shoot pole (Pence 1995). In some species, embryos are extremely sensitive to desiccation and even minor reduction in their moisture content down to levels much too high to obtain survival after freezing leads to irreparable structural damage. It should be emphasized that selecting embryos at the right developmental stage is of critical importance for the success of any cryopreservation experiment (Engelmann *et al.*, 1995) However, in these cases basic protocols for disinfection, *in vitro* germination of embryos or embryonic axes, plantlet development and possibly limited propagation will have to be established prior to any cryopreservation experiment.

Cryostorage of seed was initially developed for the preservation of genetic resources of agriculturally important species for breeding and selection. The development of comparatively simple cryostorage protocols allowed seeds of over 155 agricultural species (Stanwood, 1985) to be stored at low cost, in an environment without obvious problems of seed ageing, genetic variations and predation common to many conventional seed storage methods. With the regular use of cryostorage system for seeds of agri-crops, the same process is now viewed as having important application for preserving seeds of medicinal plants (Decruse *et al.*, 1999), endangered species (Touchel and Dixon 1994) and other native plant species (Pence 1991, Touchel and Dixon 1993, Decruse and Seeni 2002). For the long-term preservation of species producing recalcitrant seeds, zygotic embryos were used for cryopreservation. Incidentally excised zygotic embryos or embryonic axes were successfully employed for the cryopreservation of coconut (Assy-Bah and Engelman, 1992 a,b, Chin *et al.*,

1989), cocoa (Pence, 1991, Chandel *et al.*, 1995) oil palm (Chabrillange *et al.*, 1997), walnut (de Boucaud *et al.*, 1991), jack fruit (Chandel *et al.*, 1995, Thammasiri, 1999), rubber (Normann 1986), tea (Chauduryi *et al.*, 1991) and neem (Berjack and Dumet, 1996).

National Gene bank for Medicinal and Aromatic Plants at Tropical Botanic Garden and Research Institute (TBGRI) is one among the four (CIMAP, Lucknow, NBPGR, New Delhi and RRL, Jammu) having the mandate of conserving the medicinal and aromatic plants (MAPs) of Peninsular India through biotechnological intervention including collection, ex *situ* conservation and characterization of the precious taxa that are rare, endangered, threatened, endemic, vulnerable or over exploited as the case may be. TBGRI has significantly developed cryopreservation protocol on rare and endangered medicinal plants of India (Decruse *et al.*, 1999, Decruse and Seeni, 2002, Radha *et al.*, 2006). A cryobank was also established which now holds more than 25 accessions of medcinal and aromatic plants (Decruse *et al.*, 1999b, Decruse and Seeni 2002b, Radha *et al.*, 2010).

2. Cryopreservation of excised embryonic axes of *Nothapodytes nimmoniana* (Graham) Mebberly, a vulnerable medicinal tree species of the Western Ghats

Nothapodytes nimmoniana (Graham) Mebberly, of family Icacinaceae is a small vulnerable medicinal tree distributed in India, Sri Lanka, Myanmar, Thailand, Malaysia and China. In India it is distributed in upper ranges of the Western Ghats particularly in the Nilgiris and Palni hills of southern peninsula. The stem and roots are an important source of the anti-tumour quinoline alkaloid camptothecin (Hsiang etal.,1985) and also find applications against retrovirus and human immunodeficiency virus. Consequently natural population of this species in the Western Ghats are severely depleted owing to habitat destruction and over exploitation (Cragg *et al.*,1993, Ravikumar and Ved, 2000) and hence conservation efforts are undertaken by certain agencies in the region.

Seeds of N. nimmoniana are large intermediate type showed 100% germination under controlled conditions. Embryonic axes with cotyledons having moisture content of 55.7% presumed to be intermediate in nature, lose their viability within a short period after maturity. Cryopreservaton of zygotic embryos is recognized as an effective tool for the long-term preservation of such plant species those produce recalcitrant/large seeds (Engelmann, 1997).

Desiccation and cryopreservation. The seeds were separated from the fruits (drupe), rinsed in running tap water for one hour to remove the mucilage and washed in commercial detergent (1% Teepol, Godrej, India Ltd., Mumbai) for 10 min. followed by thorough washing in running tap water for 10-20 min. Seeds were then surface decontaminated by immersion in 0.01% (w/v) $HgCl_2$ for 5-10 min. followed by 3-5 rinses in sterile distilled water. Seed coat was broken and embryos with cotyledons were dissected out free of the endosperm in aseptic condition in the laminar air flow cabinet. Immediately after dissection, batches of 20 embryos each were subjected to dehydration under laminar airflow for 30, 60, 90, 120,150,180 and 210 min. period. A sample of 10 embryos was inoculated into MS medium (Murashige and Skoog, 1962) devoid of PGR as fresh control and cultured under 10/14h light/dark periods (30 - 50 μmol m^{-2} s^{-2}) at 25±2 ^{0}C for 8 weeks. After desiccation at 30 min intervals, equally divided samples of 10 embryos were transferred to germination medium and another 10 packed in 2ml cryovial and transferred to LN (at -196 ^{0}C) After 24h storage, the vials were retrieved from LN and rewarmed in a water bath at 40⁰C for 1-2 min. The rewarmed embryos were also transferred to germination medium and cultured under stated conditions for recovery. The whole experiment was repeated three times.

Observations on the germination of embryos were made after 8 weeks and results analyzed statistically in a completely randomized model. Survival rate was assessed as the percentage of embryonic axes that exhibited any kind of growth, including seedling development; shoot growth and root growth.

Moisture content determination. Moisture content (MC) of the embryos was determined by constant temperature oven method (103 ⁰C) for 17h.

The embryonic axes with cotyledons (Fig 1a) freshly dissected from the seeds possessed 55.7% MC and exhibited 86.67% germination and normal growth in MS medium devoid of PGRs within a week of culture. Dehydration under laminar airflow reduced the MC to 43.7% after 30min and 31.3% after 60min. without appreciable reduction in viability so that 76-77% of them germinated (Fig.1). Dehydration for 120min reduced MC to 19.6% and germination to 66.67% and was the optimum dehydration period (Fig.2) to get maximum germination (60%) after LN treatment (Fig.2). Root and shoot emergence was observed after one week of culture (Fig.1b) in 60% of the desiccated (120 min.) and LN treated embryonic axes and well developed seedlings were obtained within 20 days of culture (Fig.1c). Dehydration beyond 120 min. gradually reduced MC and drastically reduced viability. The MC came down to 12.1% after 210 min. when none of the embryos survived. Prolonged dehydration (150-180min) not only reduced survival down to 16.67-10% but also caused abnormal growth with only radicle development in the survived embryos (Fig.1d).

Research in the past two decades has shown that most orthodox seeds remain viable for long periods of storage after attaining appropriate desiccation levels of about 3-5% moisture content (Roberts, 1973). Contrary to this, recalcitrant seeds of several tropical and temperate species are desiccation sensitive, eg. Tea, Cocoa, Citrus, Jack fruit (Chin and Roberts, 1980). There are various options available to improve storage of non-orthodox seeds/embryos. Desiccation is the simplest procedure since it consists of dehydrating explants, and then freezing them rapidly by direct immersion in LN has been applied to embryonic axes extracted from recalcitrant and intermediate seeds (Engelmann, 1997). It should also be noted that selection of embryos at the right developmental stage is of critical importance for the success of any cryopreservation experiment (Engelmann *et al.*, 1995). The conservation efforts of *N. nimmoniana* are hampered mainly due to relatively large and intermediate type of seeds with desiccation sensitivity. The viability of the embryos was not much affected when the embryos were desiccated from 55.7% to 43.7% (i.e. 30 min. desiccation). Significant loss of viability due to further reduction of moisture content shows the intermediate nature of the embryos is in line with the report of Dussert *et al* (1995). Safe moisture content of the embryonic axes as obtained in the present study is 19.6% (60% survival). Damage to plumule rather than radicle occurred due to excessive dehydration of *N. nimmoniana* embryos is as observed earlier in *Auracaria hunstenni* where desiccation damage is reported to be more serious in the plumule (Pritchard and Prendergast, 1986). The exact causes of embryonic death and its relationship with moisture content are not fully understood. Chin *et al* stated that seed death could be due either to the moisture content falling below a critical value or simply a general physiological deterioration with time. If embryonic axes have been desiccated to around 20% moisture content without loss of viability, it is possible that

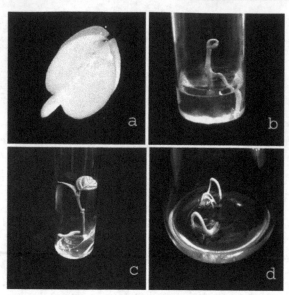

Fig..1 a. Isolated embryonic axes, b. Cryopreserved embryo showing germination after 30 days of culture on MS basal medium, c. Seedling from Cryopreserved embryo after 60 days of culture on MS basal medium and d. radicle development and degeneration of plumule in embryo subjected to desiccation for 180 min.

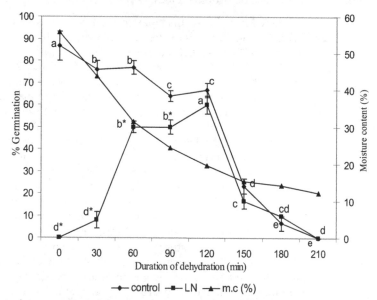

Fig. 2. Effect of cryopreservation on germination of *N. nimmoniana* zygotic embryos. Different letter (s) in a data series shows significant difference at 5% level based on LSD multiple 't' test. *Control and LN treated values differ significantly at 5% level based on Student 't' test. The bars represent SEM.

cooling and storage in LN will be progressed more easily. In most of the reports of successful cryopreservation, excised embryos or embryonic axes have been used for desiccation sensitive species, i.e. zygotic embryos of Citrus (Mumford and Grout, 1979) Oil palm (Grout *et al.*, 1983) Coconut (Chin *et al.*, 1989) Hevea (Normah *et al.*,1986) where the embryos withstand freezing after being subjected to partial desiccation. The desiccated embryonic axes do not lose viability after rapid cooling and storage at the temperature of LN. At such temperature there should be no change in the tissue either genetic or developmental, over a period of decades (Ashwood *et al.*, 1977). This situation together with the ease to develop independent plants *in vitro* (Satheeshkumar and Seeni, 2000, Ravishankar Rai, 2002) from embryonic axes suggest cryopreservation is an effective technique for the long-term conservation of *N. nimmoniana*, a medicinal tree species producing large intermediate type of seeds.

3. References

Ashwood Smith MJ and Grant E (1977). The freezing of mammalian embryos, at *CIBA symposium*, (eds) Elliot K, Whelan J, Elsevier, Holland, 251-271.

Assy-Bah B and Engelman F (1992a). *Cryo-Letters* 13: 67-74

Assy-Bah B and Engelman F (1992b). *Cryo-Letters* 13: 117-26

Bajaj YPS (1995). Cryopreservation of plant cell, tissue and organ culture for the conservation of germplasm and biodiversity. In: Bajaj YPS (ed) *Biotechnology in Agriculture and Forestry Cryopreservation of Plant Germplasm I*, New York, Springer-Verlage, pp. 3-18.

Bekheet SA, Taha HS, Saker MM, Solliman ME (2007). *J Appl Sci Res.*, 3 (9): 859-866.

Benson EE (1999). Cryopreservation. In: Benson EE (ed) *Plant Conservation Biotechnology*, Taylor and Francis, London, pp. 83-95.

Berjack P and Dumet D (1996). *Cryo-Letters* 17: 99-104

Chabrillange N, Aberlenc-Bertossi F, Engelman F and Duval Y (1997). *Cryo-Letters* 18: 68-74.

Chandel KPS, Chaudhury R, Radhamani J and Malik SK (1995). *Ann.Bot.*, 76: 443-450.

Chang Y and Reed BM (2001. *HortSci* 36 (7): 1329-1333.

Chaudury R, Radhamani J and Chandel KPS (1991). *Cryo-Letters* 12: 31-6

Chin HF, Krishnapillay B and Hor YL (1989). *Pertanika* 12: 183-86

Chin HF, Krishnapillay B and Standwood PC (1989). In: *Seed Moisture* (eds. P C Standwood and M B MC Donald). *Crop Science Society of America*, Madison, WI, USA, 15-22.

Chin HF and Pritchard HW (1988). Recalcitrant seeds, A Status Report. *IBPGR*, Rome.

Chin HF and Roberts EH (1980). *Recalcitrant crop seeds*, Tropical press SDN, BHD.

Cragg GM, Schepartz SA, Suffness M, Grever MR (1993). *J Nat. Prod.* 56:1657-1668.

De Boucaud M, Brison M, Ledoux C, Germain E and Lutz A (1991). *Cryo-Letters* 12:163-166

Decruse SW, Seeni S and Pushpangadan P (1999a) *Seed Sci and Technol.* 27: 501- 505

Decruse SW Seeni S and Pushpangadan P (1999b). *Cryo-Letters* 20: 243-250

Decruse SW and Seeni S (2002b). *Cryo-Letters* 23:55-60

Decruse SWand Seeni S (2002a). *Seed Sci and Technol* (In Press)

Engelmann F (1997). *Plant Genetic Resources Newsletter* 112: 9-18.

Engelmann F (2004). *In Vitro Cell Dev Biol Plant* 40: 427-433.

Engelmann FD, Dumet N, Chabrillange A, Abdelnour Esquivel B, Assy-Bah J Dereuddre and Duval Y (1995). *Plant Genetic Resources Newsletter* 103: 27-31.

González-Benito ME, Clavero-Ramirez I, López-Aranda JM (2004). *Spanish J Agric Res* 2 (3): 341-351.

Hsiang YH, Herzberg R, Hecht S and Liu LF (1985). *J Biol Chem* 260: 14873-14878.

Kaczmarczyk A, Shvachko N, Lupysheva Y, Hajirezaei MR, Keller ERJ (2008). *Plant Cell Rep* 27: 1551-1558.

Kartha KK, Engelmann F (1994). Cryopreservation and germplasm storage. In: Vasil, I.K. and Thorpe, T.A. (eds) *Plant Cell Tiss Cult* Springer 195-230.

Martin C, Iridono JM, Benito-Gonzales E, Perez C (1998). *Agro-Food-Ind Hi-Tech* 9 (1): 37-40.

Mumford PM, Grout BWW (1979). *Seed Science Tech* 7: 407-410.

Murashige T and Skoog F (1962). *Physiol Plant* 15: 473-497.

Normah MN, Chin HF and Hor LY (1986). *Pertanika* 9: 299-303.

Panis B, Lambardi M (2005). Status of cryopreservation technologies in plants (Crops and Forest trees).At: *The Role of Biotechnology*, Villa Gualino, Turin, Italy.

Paunesca A (2009). *Romanian Biotech Letters* 14 (1): 4095-4104.

Pence VC (1991). *Plant Cell Rep.* 10: 144-7

Pence VC (1991). *Seed Sci. and Technol.* 19: 235-251

Pence VC (1995). In: Bajaj YPS (ed) *Cryopreservation of Plant Germplasm, Biotechnology in Agriculture and Forestry*, Springer, Berlin Heidelberg, pp. 29-50.

Pritchard HW, and Prendergast, FG (1986). *J. Exp. Bot* 37: 1388-1397

Radha RK, Decruse SW, Seeni S (2006). Cryopreservation of embryonic axes of recalcitrant seed species *Myristica malabarica* Lam., a rare medicinal plant of the Southern Western Ghats. Presented in the *National Seminar on Plant Resources of the Western Ghats* organized by Karnataka Biodiversity Board at IISc. Bangalore on 7th and 8th December.

Radha RK, Decruse SW, Krishnan PN (2010). *Ind. J. Biotechnology* 9: 435-437.

Rajasekharan PE (2006). Prospects of new cryopreservation techniques for conservation oftropical horticultural species. Paper presented at the *ICAR Short Course on In Vitro Conservation and Cryopreservation-New Options to Conserve Horticultural Genetic Resources*, Banglore, India, 21-30 September.

Ravikumar K, and Ved,DK (2000). Illustrated field guide of 100 Red listed medicinal plants of Conservation concern in Southern India. Foundation for Revitalization of Local Health traditions (FRLHT), Bangalore, pp 465.

Ravishankar Rai V (2002). *In Vitro Cellular & Developmental Biology. Plant* 38 (4): 347-351.

Reed BM (2001). *Cryo-Letters* 22:97-104

Roberts EH (1973). *Seed Sci Tech*, 1: 499-514

Sakai A (2000). Development of cryopreservation techniques. In: Engelmann F Takagi H (eds) *Cryopreservation of tropical plant germplasm*. International of Plant Genetic Resources Institute, Rome, pp. 1-7.

Satheeshkumar K and Seeni S (2000). *Ind J Expt Biol* 38: 273-277.

Stanwood PC (1985). In: Kartha KK ed. *Cryopreservation of plant cells and organs* (Boca Raton, Florida: CRC press) pp 199-226

Stanwood PC (1987). *Crop Sci* 27: 327-331.

Scowcroft WR (1984). Genetic variability in tissue culture: Impact on germplasm conservation and utilization. *International Board for Plant Genetic Resources* Secretariat, Rome, p. 42.

Touchel DH and Dixon KW (1993). *Biodiversity and Conservation* 2:594-602

Touchel DH and Dixon KW (1994). *Annals of Bot.* 74: 541-546

Walters C, Volk GM, Towill LE, Forsline P (2009). Survival of cryogenically-stored dormant apple buds: a 20 year assessment. Paper presented at the *1st International Symposium on Cryopreservation in Horticultural Species, Leuven, Belgium*, 5-9 April.

Westwood MN (1989). *Plant Breeding Rev* 7: 111-128.

Zhao MA, Dhital SP, Fang YL, Khu DM, Song YS, Park EJ, Kang CW and Lim HT (2005). *J Plant Biotech* 7 (3):183-186

7

Somatic Embryogenesis and Cryopreservation in Forest Species: The Cork Oak Case Study

Conceição Santos

CESAM & Department of Biology, University of Aveiro, Aveiro
Portugal

1. Introduction

1.1 General concepts

It is widely accepted that propagation and conservation strategies are crucial in any forest breeding program and that plant biotechnology has powerful tools to propagate and preserve selected genotypes (Jain, 1999; Park et al., 1988; Park, 2002). In particular, in vitro techniques have demonstrated to be essential in a large number of agricultural and forest breeding programs, allowing large scale clonal propagation of elite genotypes (Pinto et al., 2008) or of endangered forest species (Brito et al., 2009). In vitro techniques are also essential in breeding programs involving genetic improvement (e.g. using techniques such as somatic hybridization or genetic transformation) or involving long term germplasm conservation by cryopreservation (Benson, 2008; Fernandes et al., 2008).

Somatic embryogenesis offers advantages in improving forest species over other in vitro propagation methods (Park, 2002), namely: a) somatic embryos simultaneously possess both the shoot and root meristems; so a distinct rooting stage, which usually involves stressing procedures, is not necessary; b) during somatic embryogenesis, embryos/clusters are often formed faster and potentially at extremely high numbers per explant; c) somatic embryogenesis is amenable to automation, which means it may become cheaper than other clonal propagation techniques (Pinto et al., 2002; 2008); d) finally, a robust and efficient protocol of somatic embryogenesis will allow that embryogenic clonal lines can be preserved for long periods (e.g., in liquid N_2) while corresponding plants are transferred to field conditions and are monitored for their characteristics.

The combination of somatic embryogenesis and conservation strategies in breeding programs allows that one may develop high-value forest clonal varieties merely by recovering from N_2 those genotypes/clones that showed best characteristics in the field. These selected genotypes can later be used for advanced breeding programs and commercial forestry (Park, 2002).

From the exposed advantages, it is generally accepted among breeders that breeding programs using in vitro clonal strategy require as a pre-requisite that a cloning technique (preferably somatic embryogenesis) is well established. Then it also requires the optimization of long-term genetic testing methodologies of clonal lines and long term conservation. The combination of these strategies will then enable large-scale production and disposition of tested clonal lines in industrial forest management. For example, micropropagation processes are well refined for spruce and larch species to support their commercial application (http://cfs.nrcan.gc.ca/ factsheets/conifersomatic). However, for most pine species, it is much more difficult to obtain somatic embryogenesis, though interesting advances are in course (e.g. Park et al., 2010). Similarly, the application of this technology to most forest dicotyledonous species, as is the case of *Quercus* genus, has demonstrated to be difficult due to species general recalcitrance to in vitro culture (Santos, 2008).

1.2 *Quercus suber* in vitro cloning: A reliable protocol?

Cork oak (*Quercus suber* L.) belongs to Fagaceae, an important family of forest trees in the Northern hemisphere dominating temperate forests and Mediterranean ecosystems. In particular, cork oak is an abundant species in the Atlantic and West Mediterranean countries where it is an important component of Mediterranean ecosystems (Pinto et al., 2002). Cork is the bark of the oak, which is a natural, renewable and sustainable raw material product of economic interest for a range of applications. Due to its enormous economical importance, intense research has been focused on cork valuable material and, more recently, on cork oak germplasm-conservation programs (Fernandes et al., 2008).

In the last decades, studies were done to improve protocols of cork oak in vitro micropropagation and conservation (namely cryopreservation). In particular, the currently available rates of success in cork oak plant regeneration by somatic embryogenesis (Fernandes et al., 2011; Lopes et al., 2006; Loureiro et al., 2005; Pinto et al., 2001; 2002) and in cork oak material cryopreservation (Fernandes et al., 2008; Fernandes, 2011) are highly encouraging for researchers and breeders to consider the integration of these strategies in breeding and conservation programs of this species (Figure 1).

Recently, it was also established a Portuguese consortium to identify and characterize cork oak ESTs Gene responses to several biotic and abiotic stresses as well as to developmental conditioning are currently being screened and data will be of upmost importance to cork oak researchers and breeders (http://www.fct.pt/apoios/projectos/consulta/projectos.phtml.en).

For long it has been assumed that *Quercus* species have, at some extent, recalcitrant responses to micropropagation in general, and to somatic embryogenesis in particular. Most common strategies for cork oak micropropagation use stem cuttings or use juvenile material or leaves for somatic embryogenesis (Santos, 2008; Fernandes et al., 2008). When utilizing material selected from adult field trees as explant sources, the use of greenhouse forced sprouts instead of directly collected field material is strongly advised.

The developed micropropagation by stem cutting is efficient with juvenile and, less, mature genotypes. Briefly, after disinfection with sodium hypochloride, explants (with 1-2 apical and/or lateral buds) are inoculated on WPM ("Woody Plant Medium", Lloyd and McCown, 1980) medium containing benzylaminopurine (BAP 0.5 mg/L) and naphthalene acetic acid (NAA 0.1 mg/L). After multiplication and elongation, shoots are exposed to an indol-butiric acid shock for rooting. Plants in this stage are then ready for acclimatization (Pires et al., 2003; Figure 2a).

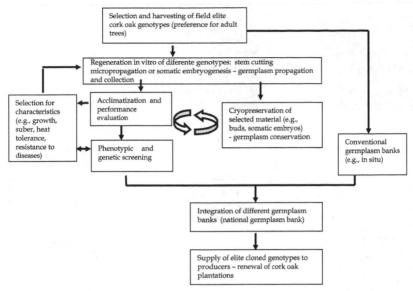

Fig. 1. Schematic representation of a proposed strategy of integrating cork oak micropropagation and cryopreservation technologies in Portuguese breeding programs of this species (adapted from Santos, 2008).

As reported above, somatic embryogenesis is a regeneration strategy with enormous potential for breeding programs. Somatic embryos were developed in *Q. canariensis* (Bueno et al., 1996; 2000), *Q. rubra* (Vengadesan & Pijut, 2009), *Q. serrata* (Ishii et al., 1999; Takur et al., 1999), *Q. robur* (Cuenca et al., 1998; Endemann et al., 2002; Wilhelm et al., 1999), *Q. acutissima* (e.g., Kim, 2000) and *Q. petrea* (Chalupa, 2005). However, not only most studies use juvenile sources of explants (e.g., zygotic embryos and seedlings), but also plant conversion frequencies are still low, supporting the recalcitrance of these species.

Q. suber somatic embryogenesis was obtained first from juvenile plants (e.g., Bueno et al., 1996; 2000; Pinto et al., 2001) and later from leaf explants of mature plants (e.g., Hernandez et al., 2003; Lopes et al., 2006; Pinto et al., 2002; Santos et al., 2007) (Figure 2b,c).

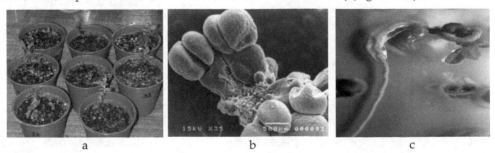

Fig. 2. Micropropagation from field mature cork oak trees: a) Acclimatized plants micropropagated by stem cuttings; b) Scanning electon microscopy of two cotyledonary somatic embryos; c) converted embling (Adapted from Pires et al., 2003; Pinto et al., 2002; Santos, 2008).

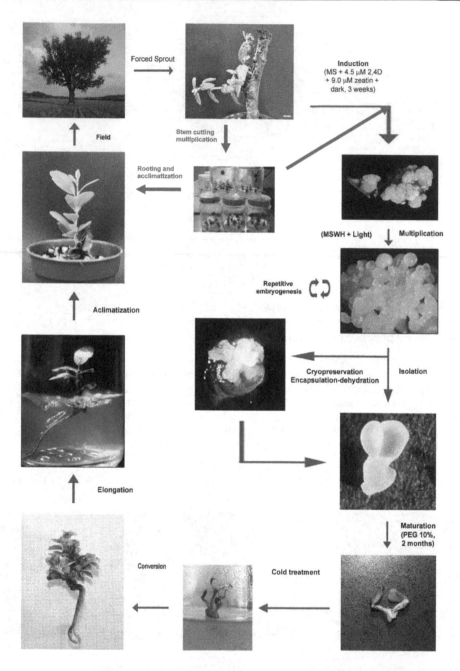

Fig. 3. Enhanced protocol of our group for cork oak somatic embryogenesis. MS medium - Murashige and Skoog, 1962; MSWH - MS medium with no growth regulators. (Adapted from Fernandes, 2011).

The initially protocol developed by Santos and collaborators (Pinto et al., 2002; Lopes et al., 2006) was, however, not sufficiently efficient for large scale propagation and for immediate transfer to industrial breeding programmes of cork oak. Meanwhile, those and other authors reported the deficient maturation of somatic embryos during somatic embryogenesis, as the main cause for low conversion rates in this species (Chalupa, 2005; Fernandes et al 2011; Hernández et al., 1999).

Hernández et al. (2003) highlighted that an adequate reserve deposition in the embryo tissues seems to be necessary for their adequate maturation. Efforts were made since then to manipulate physical conditions in order to promote the adequate accumulation of reserves (Fernandes, 2011; Santos, 2008). Fernández-Guijarro et al. (1995) reported that somatic embryos from cork oak young seedlings increased maturation under light followed by storage at 4 °C, and that controlled starvation could benefit synchronization.

Santos (2008) and Fernandes (2011) compared the accumulation profiles of carbohydrate, lipid and protein reserves during the maturation of cork oak somatic embryos and the zygotic counterparts. Assuming that the accumulation of reserves that occurs in zygotic embryos may be ideal for embryos maturation, Fernandes (2011) also compared the accumulation profiles of somatic embryos exposed to different conditions such as polyethylene glycol (PEG), abscisic acid (ABA) and cold, and defined the condition that led to an accumulation profile closer to the one of the zygotic embryos.

From those analyses, the authors proposed an improvement to the initial somatic embryogenesis protocol developed by Pinto et al. (2002) (see Figure 3). In the improved protocol clusters of somatic globular embryos are isolated and transferred to MS medium (Murashige & Skoog, 1962) with PEG. After maturation, cotyledonar embryos are transferred to MS medium and submitted to chilling (4° C). Conversion is then achieved on woody plant medium (WPM) medium supplemented with BAP 0.5 mg/L and NAA 0.1 mg/L. After some weeks plants are acclimatized with success (Fernandes, 2011).

In conclusion, the inclusion of cold and osmotic stress in the protocol improved somatic embryos maturation and consequent conversion in approximately 70% of the genotypes. However, it was evident a genotype dependence in this process, with responsiveness ranging from very-good/in most genotypes to null, in few genotypes (Fernandes, 2011) (Figure 3).

1.3 Current challenges for the SE process

The loss of embryogenic competence is one of the major drawbacks of long term micropropagation potocols (e.g., Brito et al., 2009). In particular, embryogenic masses were maintained for long periods may dedifferentiate and lose their embryogenic potential. In cork oak this phenomena has originated two types of calluses under the same conditions: embryogenic (EC) and non-embryogenic (NEC) and these last calluses rarely regain embryogenic ability (Santos, 2008).

In vitro functional changes during embryogenesis imply changes in explant cells from differentiated and quiescent (G_0) stage to dedifferentiated dividing (G_1-S-G_2/M) stages, and later an evolution to embryogenic states. All these transitions imply changes in gene expression, and in cell cycle dynamics, where growth regulators, namely auxins and cytokinins, are crucial players (Gahan, 2007).

Using embryogenic and non embryogenic calluses of adult cork oak genotypes, our group reported differential distribution of cells staged in G_1, S and G_2 phases according to callus and growth regulators type (Fernandes, 2011) confirming that cell cycle dynamics during somatic embryogenesis suffers exogenously-induced alterations, and in particular, it is conditioned by growth regulators (Gahan, 2007). We also found that using the two different somatic embryogenesis protocols available (Fernandes, 2011; Pinto et al., 2002), not only cell cycle dynamics changed with time during the process, but also genotypes with different somatic embryogenic competences had different cell cycle dynamics (Fernandes, 2011). Curiously, responsive genotypes showed cell cycles with similar progression profiles (Fernandes, 2011).

Considering the key players regulating cell cycle dynamics, cyclins are among the most important. In a broad sense, D-type cyclins are thought to regulate the G_1-to-S transition, A-type cyclins, the S-to-M phase control, and B-type cyclins regulate both the G_2-to-M transition and intra-M-phase control (Gahan, 2007). The cyclin D (CYCD)/retinoblastoma pathway is believed to be involved in controlling both the commitment of cells to the mitotic cell cycle and decisions involving cell growth, differentiation, and cell cycle exit (Dissmeyer et al., 2009; Cools & Veylder, 2009). Key genes for growth and cell division are regulated by E2F transcription factors, which are inactive when bound by retinoblastoma. The phosphorylation of retinoblastoma is initiated by CYCD-containing cyclin-dependent kinases (CDKs) and is completed by cyclin E–CDK2, resulting in the dissociation of retinoblastoma from E2F factors, triggering the passage of cells from G_1- to S-phase (Gahan, 2007). This key role for the G_1 exit pathway results in it being the primary and predominant cell cycle control point. However, cyclin E–CDK2 is rate-limiting for entry into S-phase and can trigger S-phase in the absence of RB phosphorylation. CYCD3;1 are the best studied examples and expression of their genes is regulated by extrinsic signals, such as sucrose availability. CYCD3;1 expression is also regulated by plant hormones (Dewitte et al., 2003; for review see Dewitte & Murray, 2003).

The E2F family plays a critical role in organizing cell cycle progression by coordinating early cell cycle events with the transcription of genes required for entry into S-phase (e.g., Inzé, 2000). Two major classes of genes possess characteristic E2F binding sites, the first class encodes essential enzymes in the pathways for nucleotide and DNA synthesis that are co-ordinately up-regulated in late G_1. The second class corresponds to genes for regulators of cell cycle progression. Genes from both classes respond to ectopic expression of E2Fs from the first sub-group, namely those that can induce entry into S-phase (Dewitte & Murray, 2003; Gahan et al., 2007).

The cell cycle involves a complex network of regulating molecules. So the control of all these classes of checkpoints regulators is under study in cork oak embryogenic (EC) and non-embryogenic (NEC) tissues (Santos, 2011, unpublished data). Understanding and controlling these checkpoints will become a powerful tool to both better understand the embryogenic per se and to manipulate the developmental stages of embryogenic process.

2. Cryopreservation

2.1 General principles

Public and private efforts have been made to protect and conserve germplasm by preserving the genetic material of selected genotypes (e.g., Engelmann, 2000). As in other species, cork oak germplasm preservation can be done in situ (in the field and in natural environment), which demands large areas and is susceptible to environmental hazards. Alternatively preservation may be done ex situ (for general review see Li & Pritchard et al., 2009). In particular, in vitro preservation allows that in a small area, large amounts of genotypes are multiplied and maintained under controlled conditions where environmental influences are minimal. However, precocious ageing as well as somaclonal variation and genetic instability may arise after long term culture (Brito et al., 2009).

Alternatively, cryopreservation is the storage of living materials at extremely low temperatures using usually liquid nitrogen (−196 °C), and is an ideal strategy for plant germplasm preservation (Benson, 2008; Feng et al., 2011; Wang & Perl, 2006). This preservation strategy allows not only the preservation of material in small volumes (involving low maintenance requirements) but also, by reducing to residual values the cell metabolism, it allows that cells are stored for long periods, with low probability of genetic instability occurrence (Feng et al., 2011). Cryopreservation therefore allows: the conservation of plant material minimizing occurrences of genetic instability, contaminations and diseases; the preservation of endangered, rare or selected genotypes. Cryopreservation is already being applied to several plant species including forest woody species (e.g., Sakai et al., 2008). Also different plant materials have been used in this preservation strategy: shoot tips, cell cultures, embryos and seeds (Feng et al., 2011).

For cryopreservation to be useful in breeding programs, it is necessary to develop the cryogenic technique *per se*, and to ensure that robust and efficient regeneration protocols are available. Freezing and thawing stages require that cells are structurally and functionally cryoprotected. This may happen naturally (e.g. some naturally dehydrated material) but usually it is induced artificially with treatment with cryoprotectants that influence ice formation and activity of electrolytes present in the solution. Ideally, cryoprotectants should have low or no cytotoxicity. Cryoprotectants may be: a) permeating compounds, such as dimethylsulphoxide (DMSO, used usually in the range of 5-10%) that has a rapid entrance rate and so requires short incubation periods; another permeating compound is glycerol (used often in the range of 10-20%); b) non-permeating compounds such as sugars, sugar alcohols, polyethylene glycol (PEG). Often mixtures of cryoprotectant compounds are preferred to improve their efficacy (e.g., combinations of PEG : glucose : DMSO). Finally other strategies as cold hardening or ABA treatment may increase the freezing resistance and survival rates of cells.

Plant cryopreservation strategies may include slow or rapid freezing approaches. The first is based on physico-chemical changes during the process, namely associated with apoplastic ice crystal formation while cytoplasm may remain free from intra-cellular ice formation. Slow freezing decreases therefore the osmotic potential of the cytoplasm contributing to the cell desiccation. Rapid freezing is achieved by immersion of the cryoprotectant-treated samples in N_2, leading to an ultra-fast cooling that prevents the formation of ice crystals inside the cell (e.g., Sakai et al., 2008).

Some variants involve vitrification that includes a cell dehydration step prior to storage in N_2 (Sakai et al., 2008). This may rely on the ability of concentrated solutions of cryoprotectants to become viscous to very low temperatures, without ice formation. Consequently, during the vitrification process plant cells are dehydrated and the cytoplasm vitrified during freezing, which allows that ice crystals are rarely formed. Vitrification has already been applied to a large number of species (Panis et al., 2005; Sakai et al., 2008).

Nonetheless, during this process, usually complex and toxic solutions with high osmotic potential, such as the PVS2, are used for cryoprotection (Fernandes et al., 2008). Also the duration of the successive steps of a vitrification protocol is in general very short, hampering the simultaneous treatment of a large number of samples.

Some alternatives to vitrification-based techniques were developed, namely encapsulation-dehydration strategies (e.g., Engelmann et al., 2008). Encapsulation-dehydration is based on concepts related with artificial seeds. Concisely, plant tissues, such as shoot tips or somatic embryos, are covered by for example alginate. Then, they are dehydrated (using exposures to highly concentrated solutions, and/or to air in a flow chamber), before being transferred to N_2. This strategy has the advantage of using less toxic compounds such as glycerol than in other vitrification methods, thus minimizing stress conditions (e.g. Volk et al., 2006). It also has the advantage of easy and inexpensive manipulation, not requiring expensive instruments, as occurs in controlled freezing (Fernandes et al., 2008).

This method has been applied to many species, such as, for example, mulberry (Niino & Sakai, 1992), *Prunus* sp. (Brison et al., 1992), sweet potato (Feng et al., 2011; Hirai & Sakai, 2003), persimmon (Matsumoto et al., 2001), apple (Niino & Sakai, 1992; Paul et al., 2000), lily (Bouman & Klerk, 1990) and even grapevine embryogenic cell suspensions (Wang & Perl, 2006) or pear (Niino & Sakai, 1992; Scottez et al., 1992). In several assays as in those with *Robinia pseudoacacia*, it was demonstrated that encapsulation-dehydration originated better results than vitrification (Verleysen et al., 2005). Recently we have also demonstrated that encapsulation-dehydration was the most efficient method of cryopreservation of *Quercus suber* somatic embryos (Fernandes et al., 2008).

2.2 *Quercus suber* cryopreservation

Propagation of cork oak presents several drawbacks as it has a high heterozigocity, often leading to individuals with high probability of instability and genetically distinct from parents, therefore leading to high numbers of undesired genotypes (Lopes et al., 2006). Moreover, seeds are only stored for short periods as they rapidly loose viability. This recalcitrance jeopardises the development of conservation and improvement programs in this species. As in most forest species, *Quercus suber* conservation approaches consist mainly in agro-forest sustainable systems, and scarce strategies using biotechnological approaches have already succeeded. Valladares et al. (2004) highlighted that highly interesting individuals may be maintained with vegetative propagation.

The cryopreservation of seeds or embryos seems therefore to have huge potential as an innovative preservation strategy, in particular in species with recalcitrance. González-Benito et al. (2002) examined different factors included in the cryopreservation protocols for *Quercus ilex* and *Q. suber* embryonic axes. The authors demonstrated that temperature of in vitro incubation played an important role, mostly for *Q. ilex* axes. *Q. suber* axes were sensitive to desiccation and cooling.

With respect to *Quercus* sp. somatic embryos, Martinez et al. (2003) and Valladares et al. (2004) successfully cryopreserved embryogenic cultures of *Q. robur* and *Q. suber*, using the vitrification method. As reported above, highly toxic cryoprotectants are used in most classical vitrification processes. To overcome these negative effects, recently our group (Fernandes et al., 2008) used a less toxic variation to the classical vitrification technique, called the encapsulation-dehydration method, to cryopreserve *Q. suber* material.

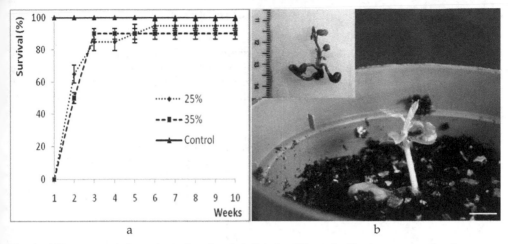

a   b

Fig. 4. a) Percentage of survival of cork oak cells after 10 weeks (for two protocols of cryopreservation used CRY25 and CRY35, corresponding to the final % content of water). b) Acclimatized plant obtained from cryopreserved somatic embryos that were recovered and cultivated on MSWH for maturation and conversion (adapted from Fernandes et al., 2008).

In this standard protocol for cork oak, somatic embryos or embryogenic clusters derived from mature trees and previously maintained on MS medium without growth regulators (MSWH) are used as samples. For encapsulation, embryos/clusters were separated and loaded in the alginate plus $CaCl_2$ solutions, forming the beads (3–4 mm), each one contained one embryo/cluster. They were then pre-cultivated on sucrose-enriched standard liquid medium (MS_{WH} with 0.7M sucrose) for 3 days. Beads were then desiccated by drying in the airflow of a laminar flow cabinet, and carefully weight loss was monitored for water content calculation. Two final water content (WC) values were assessed: 25% (CRY25) and 35% (CRY35). Afterwards, beads were placed in cryotubes (10 per vial), and immersed in N_2 for 24 h. Samples' thawing was done by incubating the cvryotubes at 38 °C (2 min) and incubating in detoxification solution (1 h). Beads were transferred to solid standard medium (MS_{WH}) for regeneration (Figure 4a, b; Fernandes et al., 2008).

This cryopreservation technique developed by Fernandes et al. (2008) for cork oak somatic embryos is simple, effective and non-toxic for the species. Also, survival rates in encapsulated-dehydrated (but non-frozen) cork oak samples achieved 90%. We also demonstrated that cryopreserved somatic embryo derived clones were able to recovery, leading to plants morphologically normal and that had genetic stability.

3. Screening of genetic variation in cloned and cryopreserved material

In vitro regenerated plants may exhibit somaclonal variation as a result from genetic or epigenetic modifications (Fourré et al., 1997; Isabel et al., 1996). It is generally accepted that morphological, cytological and molecular variations may be generated by the imposed stress during in vitro or cryopreservation processes. These induced variations are conditioned by the genotypes used and/or by the techniques/protocols used. Theoretically, any protocol of plant cloning or germplasm conservation should lead to no somaclonal variation.

Genetic variability of in vitro material or of cryopreserved samples may be assessed by several techniques. Molecular and genetic techniques when used individually give a limited perspective of the occurrence of somaclonal variation, but if combined may provide an interesting and complementary toolbox of markers for "true-to-typeness" evaluation (e.g., Santos et al., 2007).

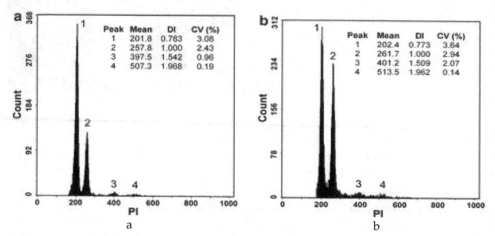

Fig. 5. Flow cytometry histograms of relative propidium iodide (PI) fluorescence intensity obtained after simultaneous analysis of nuclei isolated from *Quercus suber* and from the reference standard *Glycine max* (2C=2.50 pg DNA): **a)** leaves of the mother plant; **b)** somatic embryo. Histogram *peak 1:* nuclei at G_0/G_1 phase of *Q. suber*; *peak 2:* nuclei at G_0/G_1 phase of *G. max*, *peak 3* nuclei at G_2 phase of *Q. suber*, *peak 4* nuclei at G_2 phase of *G. max* (adapted from Loureiro et al., 2005).

Gross genetic variations of genomic origin affect mostly the number of chromosomes and ploidy level and can be detected by chromosome counting or flow cytometry (Loureiro et al., 2005). Chromosome mutations (e.g., inversion, translocation) and genic mutations are screened by molecular markers that detect DNA sequence modifications. Among these, the most popular are RFLPs (restriction fragment length polymorphisms), RAPDs (randomly amplified polymorphic DNAs), AFLPs (amplified fragment length polymorphisms) or microsatellites/SSRs (simple sequence repeats). RFLPs and AFLPs are highly reproducible techniques but are costly and may be time-consuming. For easy application of SSRs, it is required that the microsatellite loci and its flanking primers are readily available for a given species, while RAPDs is simple but may have a lack of reproducibility.

Several genetic and molecular markers have been used to assess somaclonal variation in micropropagated oak plants. Loureiro et al. (2005), using flow cytometry, found no ploidy or DNA content variations in cork oak embryogenic tissues or among somatic embryos (Figure 5).

Also, Fernandes et al. (2008) using cryopreserved material, confirmed by flow cytometric analyses that the two cryopreservation procedures (CRY25 and CRY35) provided genetic fidelity, for the parameters used: in all samples, both ploidy and DNA content were in concordance with literature data: 2C= 1.90 pg DNA (Loureiro et al., 2005; Santos et al., 2007), and changes in DNA content of non-cryopreserved and cryopreserved samples were minimal ($\leq$ 0.01 pg/2C) (Figure 6; Table 1).

Flow cytometric techniques, despite highly rapid and robust, may however not detect minor changes in DNA content, so it must be emphasised that the putative occurrence of small changes in DNA content in these kinds of assays should not be discarded.

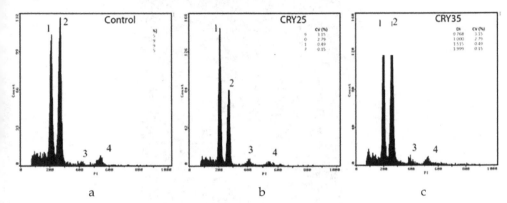

a   b   c

Fig. 5. Flow cytometric estimation of genome size of *Q. suber* samples of control (a) and after cryopreservation: CRY25 (b) and CRY35 (c). In all graphics peaks 1 and 3 correspond to 2C and 4C nuclei of cork oak somatic embryos, while peaks 2 and 4 correspond to the internal soy standard (adapted from Fernandes et al., 2008). PI: intensity of propidium fluorescence.

Tissue	DI	SD	2c DNA (pg)	SD	1c DNA (Mbp)[b]
Fresh	0.774[a]	0.0059	1.93	0.015	946
25% Cry[c]	0.777[a]	0.0073	1.94	0.018	950
35% Cry[c]	0.772[a]	0.0138	1.93	0.034	944

Table 1. Nuclear DNA content of *Quercus suber* L. fresh and cryopreserved embryos. The results are given as mean and standard deviation (SD) of the 2C DNA content in mass values (pg). Nuclear DNA content in Mbp is also given. [a] Mean values followed by the same letter are not significantly different according to the Tukey-Kramer multiple comparison test at P $\leq$ 0.01. Note: [b]1 pg DNA = 978 Mbp. [c]Cryopreserved embryos (25% and 35 % water content) (adapted from Fernandes et al 2008).

Both works supported therefore that the protocols for somatic embryogenesis and cryopreservation developed so far for cork oak led to "true-to-typeness" (for ploidy and DNA amount parameters) and were worthy of use in cork oak breeding programs.

As explored above, molecular markers provide information on sequence mutation. This information is therefore complementary to the information provided by flow cytometry (or even by chromosome counting).

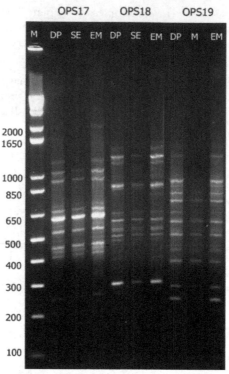

Fig. 6. DNA profiles generated by the RAPD primers OPS 17, 18 and 19, in the three different stages of the somatic embryogenesis process: donor plant (DP), somatic embryo (SE) and embling (EM). M, size marker (1Kb Plus DNA Ladder) (adapted from Fernandes et al., 2011).

In micropropagated *Q. serrate* no aberrations in the banding patterns were detected by RAPD markers (Thakur et al., 1999). Also, RAPDS were used to assess putative occurrence somaclonal variation in *Q. suber* embryogenic lines, but no molecular changes were found (Gallego et al., 1997, Sanchez et al., 2003). More recently, RAPDS were also used to evaluate genetic instability of somatic embryos of *Q. suber* obtained by the above described protocols (Figure 6; Fernandes et al., 2011).

Techniques for RAPD analyses however pose several problems of reproducibility, and give restricted information. So, other molecular analyses can provide complementary information to RAPDS. Microsatellites and AFLP are among the most used markers in *Quercus*. AFLP markers detected changes in cork oak embryogenic lines (Hornero et al., 2001). We also used AFLP to test putative genetic instability during the developed cryopreservation/somatic embryogenesis processes of cork oak (Fernandes et al., 2008). It was used six primer sets that revealed an overall high proximity value between the two vitrification-encapsulation cryopreserved (CRY25 and CRY35) and control samples. Occasionally, few extra AFLP-bands in CRY25 samples were detected and the occurrence of putative small mutations, or DNA methylation or even to subpopulation cryo-selection should not be excluded (Fernandes et al., 2008).

Locus	Allele size (bp)			
	QsG0	QsG5	QsGM1	QsGM2
QM58TGT	201	185/211	203/210	203/210
QM50-3M	276/286	285/287/292	278	282/288
QpZAG9	223/238	223/233	218/223	223/233
QpZAG15	108/123	106/120	nd	nd
QpZAG36	209/219	216/220	207/209	207/209
QpZAG110	223/233	220/238	221	221/233
QrZAG7	106/119	115/127	106/121	106/121
QrZAG11	229/263	261/273	nd	nd

Table 2. Characteristics of the microsatellite loci amplified in *Q. suber*. Allele size found in this study and allele size range and number of alleles (in parenthesis) found in other publications are also given: values for QrZAG7 and QrZAG11 are from Hornero et al. (2001*b*), values for the 6 remaining loci are from Lopes et al. (2006).

Finally, Wilhelm et al. (2005) found that that during somatic embryogenesis process of *Q. robur*, genetic instability occurred. Also, Lopes et al. (2006) confirmed genetic stability of somatic embryos and emblings derived from our somatic embryogenesis protocol (Lopes et al., 2006; Pinto et al., 2002) and after, Fernandes et al. (2008) also confirmed this stability using SSRs in material after recovering from cryopreservation (Table 2). Together with this overall genetic stability, the regenerated cork oak plants looked normal, healthy and well developed shoot. The authors concluded that the encapsulation-dehydration cryopreservation protocol used in cork oak somatic embryos was an efficient method of storage, regarding several parameters: recovery and survival rates and genetic/morphologic stability. To support the battery of protocols for molecular and genetic analyses of cork oak, Santos et al. (2007) published in detail the reliable protocol for analysis of cork oak material by SSRs and ploidy/nDNA quantification, where technical aspects and potential troubleshooting that may occur during analysis of this material are deeply discussed.

4. Concluding remarks

The utility of plant biotechnology tools in woody forest species propagation and preservations has been recognised decades ago, but only recently, it has been effectively incorporated in industrial breeding programs. Despite no robust and efficient protocol for cork oak regeneration by somatic embryogenesis is available yet, the advances observed in the last decade, together with the already available protocol for cryopreservation of this species, open perspectives for the incorporation of these two approaches in future breeding program of this species. Moreover, with the available protocols stable genotypes were obtained.

Genetic and molecular stability was assessed using complementary genetic and molecular techniques such as flow cytometry, RAPDS, AFLP and SSRs. Finally, an interesting research field will focus on the control of cell cycle progression order to control different stages of somatic embryogenesis and preservation. It is our believe that by manipulating proteins that control cell cycle phases transition (which are at the basis of differentiated/ undifferentiated cells) we'll be able to manipulate the reversion phenomena between NEC and EC and also to better control the developmental somatic embryogenesis stages.

5. References

Benson E (2008) Cryopreservation Theory. In: Plant Cryopreservation – A practical guide (B Reed, Ed.) Springer Science, ISBN 978-0-387-72275-7, Corvallis, OR, USA pp. 15-32

Bouman H & G Klerk (1990) Cryopreservation of lily meristems. Acta Horticulturae. 266: 331-337

Brison M, V Paulus, T Boucaud & F Dosba (1992) Cryopreservation of walnut and plum shoot tips. Cryobiology 29: 738-742

Brito G, A Costa, C Coelho & C Santos (2009) Large scale field acclimatization of Olea maderensis micropropagated plants: morphological and physiological survey. Trees 23: 1019-1031

Bueno M, A Gómez & J Manzanera (2000) Somatic and gametic embryogenesis in Quercus suber L. In: Somatic embryogenesis in woody plants (Jain S, Gupta P & Newton R, Eds) Kluwer Academic Publishers, ISBN 0-7923-3070-6, Netherlands, pp 479-508

Bueno M, A Gómez, O Vicente & J Manzanera (1996) Stability in ploidy level during somatic embryogenesis in Quercus canariensis. In: Somatic Cell Genetics and Molecular Genetics of Trees (Ajuha M, W Bourjan & Neale D, Eds). Kluwer Academic Publishers, Dordrecht, Netherlands, pp 23-28

Chalupa V (2005) Plant regeneration by somatic embryogenesis from cultured immature embryos of oak (Quercus robur L.) and linden (Tilia cordata Mill.) Plant Cell Reports 9: 398-401

Cools T & L Veylder (2009) DNA stress checkpoint control and plant development. Current Opinion in Plant Biology 12 (1): 23-28

Cuenca B, M San-José, M Martínez, A Ballester & A Vieitez (1998) Somatic embryogenesis from stem and leaf explants of Quercus robur L. Plant Cell Reports 18 (7-8): 538-543

Dewitte W, C Khamlichi, S Scofield, J Healy, A Jacqmard, Kilby & J Murray (2003) Altered Cell Cycle Distribution, Hyperplasia, and Inhibited Differentiation in Arabidopsis Caused by the D-Type Cyclin CYCD3. Plant Cell 15(1): 79–92

Dewitte W & J Murray (2003) The Plant Cell Cycle. Annu. Rev. Plant Biology 54: 235–64

Dissmeyer N, A Weimer, A Pusch, K Schutter, C Kamei, N Nowack, G Dua, Zhu Y, De Veylder L & A Schnittger (2009) Control of Cell Proliferation, Organ Growth and DNA Damage Response Operate Independently of Dephosphorylation of the Arabidopsis Cdk1 Homolog CDKA;1. The Plant Cell 21: 3641-3654.

Endemann M, K Hristoforoglu, T Stauber & E Wilhelm (2002) Assessment of age-related polyploidy in Quercus robur L. somatic embryos and regenerated plants using DNA flow cytometry. Biologia Plantarum, 44(3): 339-345.

Engelmann F, M Arnao, Y Wu & Escobar (2008) Development of encapsulation dehydration. In: Plant Cryopreservation – A practical guide (B Reed, Ed.) Springer Science, Corvallis OR USA, ISBN 978-0-387-72275-7, pp 59-76

Engelmann F (2000) Importance of cryopreservation for the conservation of plant genetic resources. In: Cryopreservation of tropical germplasm. Current research progress and application (Engelmann F & Takagi H, Eds) JIRCAS, ISBN Rome, Italy pp. 8-20

Feng C, Z Yin, Y Ma, Z Zhang, L Chen, B Wang, B Li, Y Huang & Q Wang (2011) Cryopreservation of sweetpotato (Ipomoea batatas) and its pathogen eradication by cryotherapy. Biotechnology Advances 29: 84–93

Fernandes P, C Rocha, A Costa & C Santos (2011) genetic stability evaluation of quercus suber l. Somatic embryogenesis by RAPD analysis. Pakistan Journal Botany, in press.

Fernandes P, E Rodriguez, G Pinto, I Roldán-Ruiz, M DeLoose & C Santos (2008) Cryopreservation of Quercus suber somatic embryos by encapsulation-dehydration and evaluation of genetic stability. Tree Physiology 28: 1841-1850.

Fernandes P (2011) Strategies of Quercus suber micropropagation. PhD Thesis, University Aveiro, Aveiro Portugal, pp 155

Fernández-Guijarro B, C Celestino & M Toribio (1995) Influence of external factors on secondary embryogenesis and germination in somatic embryos from leaves of Quercus suber Plant Cell, Tissue and Organ Culture 41 (2): 99-106

Fourré J, P Berger, L Niquet & P Andre (1997) Somatic embryogenesis and somaclonal variation in Norway spruce: Morphogenetic, cytogenetic and molecular approaches. Theoretical Applied Genetics 94(2): 159-169

Gallego F, I Martinez, C Celestino & M Toribio (1997) Testing somaclonal variation using RAPD's in Quercus suber L. somatic embryos. International Journal Plant Sciences 158(5): 563-567

Gahnn P (2007) Totipotency in the cell cycle. In: protocols for micropropagation of woody trees and fruits. (M Jain & H Haggman Ed) Springer, Dordrecht Netherlands. Pp 3-15.

Gonzalez-Benito ME, RM Prieto, E Ilerradon & C Martin (2002) Cryopreservation of Quercus suber and Quercus ilex embryonic axes: In vitro culture, desiccation and cooling factors. Cryoletters 23:283-290

Hernández I, C Celestino & M Toribio (2003) Vegetative propagation of Quercus suber L. by somatic embryogenesis. I. Factors affecting the induction in leaves from mature cork oak trees. Plant Cell Rep. 21: 759- 64

Hernández I, C Celestino, J Martinez, J Hornero, J Gallego & M Toribio (1999) Induction of somatic embryogenesis in leaves from mature Quercus suber trees. Paper presented at the Physiology and control of plant propagation in vitro. Report of activities, Eur Cost Action 822, Krakow, Poland.

Hirai D & A Sakai (2002) Simplified cryopreservation of sweet potato [Ipomoea batatas (L.) Lam.] by optimizing conditions for osmoprotection Plant Cell ReportsVolume 21, Number 10

Hornero J, I Martinez, C Celestino, FJ Gallego, V Torres & M Toribio (2001) Early checking of genetic stability of cork oak somatic embryos by AFLP analysis. Int. J. Pl. Sci., 162: 827-833.

Inzé D (2000) (ed) The plant cell cycle Klluwer Academic Press. Dordrecht, Netherlands

Isabel N, R Boivin, C Levasseur, PM Charest, J Bousquet & FM Tremblay (1996) Occurrence of somaclonal variation among somatic embryo-derived white spruces (Picea glauca, Pinaceae). Am J Bot 83(9):1121-1130

K Ishii, R Thakur & S Jain (1999) Somatic embryogenesis and evaluation of variability in somatic seedlings of Quercus serrata by RAPD markers. In: Somatic embryogenesis in woody plants (Mohan Jain S, Gupta P & Newton R, Eds), Kluwer Academic Publishers, Dordrecht, Netherlands pp 403-414

Jain SM (1999) An overview on progress of somatic embryogenesis in forest trees. In: Plant Biotechnology and In Vitro Biology in the 21st Century (Altman A, Ziv M & Izhar S, Eds), Kluwer Academic Publishers, Dordrecht, The Netherlands. pp. 57–63

Kim Y (2000) Somatic embryogenesis in *Quercus acutissima*. In: Somatic embryogenesis in woody plants (Jain S, Gupta P & Newton R, Eds). Kluwer Academic Publishers, Dordrecht, Netherlands pp 671-686

Li Z & H Pritchard (2009) The science and economics of ex situ plant conservation. Plant Cell 14(11) 614-622

Lloyd G & B McCown (1981) Commercially-feasible micropropagation of Mountain Laurel, Kalmia latifolia, by shoot tip culture. Proceeding International Plant Prop. Soc. 30: 421-427

Lopes T, G Pinto, J Loureiro, A Costa & C Santos (2006) Determination of genetic stability in long-term somatic embryogenic cultures and derived plantlets of cork oak using microsatellite markers. Tree Physiology 26: 1145-1152

Loureiro J, G Pinto, T Lopes, J Doležel & C Santos (2005) Assessment of ploidy stability of the somatic embryogenesis process in *Quercus suber* L. using flow cytometry. Planta 221(6): 815-822

Martinez M, A Ballester & A Vieitez (2003) Cryopreservation of embryogenic cultures of Quercus robur using desiccation and vitrification procedures. Cryobiology 46: 182-189

Matsumoto T, K Mochida, H Itamura & A Sakai (2001) Cryopreservation of persimmon (Diospyros kaki Thunb.) by vitrification of dormant shoot tips. Plant Cell Reports 20: 398-402

Murashige T & F Skoog (1962) A revised medium for rapid growth and bio assays with tobacco tissue cultures. Physiologia Plantarum 15: 473-497

Niino T & A Sakai (1992) Cryopreservation of alginate-coated in vitro-grown shoot tips of apple, pear and mulberry Plant Science 87: 199-206

Panis B & M Lambardi (2005) Status of cryopreservation technologies in plants (crops and forest trees). FAO: The role of biotechnology for the characterisation and conservation of crop, forest, animal and fishery genetic resources in developing countries.

Park Y (2002) Implementation of conifer somatic embryogenesis in clonal forestry: technical requirements and deployment considerations Ann. Forest Sci. 59: 651-656

Park Y, J Barrett & J Bonga (1998) Application of somatic embryogenesis in high-value clonal forestry: deployment, genetic control, and stability of cryopreserved clones. In Vitro Cell Dev Biol - Plant 34: 231–239.

Park Y, K Klimaszewska , J Park & S Mansfield (2010) Lodgepole pine: the first evidence of seed-based somatic embryogenesis and the expression of embryogenesis marker genes in shoot bud cultures of adult trees. Tree Physiol. 30(11):1469-78.

Paul H, G Daigny & B Sangwan-Noreel (2000) Cryopreservation of apple (Malus x domestica Borkh.) shoot tips following encapsulation-dehydration or encapsulation-vitrification. Plant Cell Reports, 19, 768-774

Pinto G, Y-S Park, S Silva, C Araújo, L Neves & C Santos (2008) Factors affecting maintenance, proliferation, and germination of secondary somatic embryos of Eucalyptus globulus Labill. Plant Cell Tissue and Organ Culture 95(1):69-78.

Pinto G, R Amaral, C Santos & O Carnide (2001) Somatic embryogenesis in calluses of leaves from three year old *Quercus suber* L. plants. In Meeting of the Cost 843 WG3 – II. *Quality enhancement of plant production through tissue culture*, Carcavelos, Portugal.

Pinto G, H Valentim, A Costa, S Castro, C Santos (2002) Somatic embryogenesis in leaf callus from a mature *Quercus suber* L. tree. In Vitro Cellular & Developmental Biology-Plant 38(6): 569-572

Pires I, G Pinto, J Loureiro, C Santos (2003) Study of anatomical and physiological changes in vitro cork oak plants during acclimation. XXXVIII Annual Meeting of the Society of Electron Microscopy and Cell Biology. Ponta Delgada, Portugal (5-7 de Dec) pp 29

Sakai A, D Hirai & T Niino (2008) Development of PVS-based vitrification and encapsulation-vitrification protocols. In Plant Cryopreservation – A practical guide (B Reed Ed.) Springer Science, Corvallis, OR, USA, ISBN 978-0-387-72275-7, pp 33-58

Sanchez M, M Martinez, S Valladares & A Vieitez (2003). Maturation and germination of oak somatic embryos originated from leaf and stem explants: RAPD markers for genetic analysis of regenerants. Journal of Plant Physiology 160: 699-707

Santos C (2008) De que forma é que a embriogénese somática em lenhosas é condicionada pela morte cellular. POCI/ AGR/ 60672/2004. MCT-FC -Final Report, Portugal, 59 pp

Santos C, J Loureiro, T Lopes & G Pinto (2007) Genetic fidelity analyses of in vitro propagated cork oak (*Quercus suber* L.) In: Protocols for micropropagation of woody and fruit trees (Mohan Jain S., Häggman H. Eds) Springer, ISBN 978-1-4020-6351-0, Dordrecht, The Netherlands. pp 67-84

Scottez C, E Chevreau, N Godard , Y Arnaud, M Duron & J Dereuddre (1992) Cryopreservation of cold-acclimated shoot tips of pear in vitro cultures after encapsulation-dehydration. Cryobiology 29 (6): 691-700

Thakur R, S Gota, K Ishii & S Jain (1999) Monitoring genetic stability in *Quercus serrata* Thunb. somatic embryogenesis using RAPD markers. Journal of Forestry Research, 4: 157-160

Valladares S, M Toribio, C Celestino, A Vieitez (2004) Cryopreservation of embryogenic cultures from mature *Quercus suber* trees using vitrification. Cryoletters 25: 177-186

Vengadesan G & P Pijut (2009) Somatic embryogenesis and plant regeneration of northern oak (*Quercus rubra*). Plant cell Tissue Organ culture 97(4): 1-9

Verleysen H, G Samym, E Van Bockstaele & P Debergh (2004) Evaluation of analytical techniques to predict viability after cryopreservation. Plant Cell Tissue and Organ Culture 77: 11-21

Volk G (2010) Application of functional genomics and proteomics to plant rryopreservation. Current Genomics 11: 24-29

Wang Q & A Perl (2006) Cryopreservation of embryogenic cell suspensions by encapsulation-vitrification. Methods Molecular Biololy 318: 77-86

Wilhelm E, M Endemann, K Hristoforoglu, C Prewein & M Tutkova (1999) Somatic embryogenesis in oak (*Quercus robur* L.) and production of artificial seeds. In: Proceedings of application of biotechnology to forest genetics. Biofor 99, Vitoria-Gasteiz

Wilhelm E, K Hristoforoglu, S Fluch & K Burg (2005) Detection of microsatellite instability during somatic embryogenesis of oak. Plant Cell Reports, 23: 790-795

8

Cryopreservation of Tropical Plant Germplasm with Vegetative Propagation – Review of Sugarcane (*Saccharum* spp.) and Pineapple (*Ananas comusus* (L.) Merrill) Cases

Marcos Edel Martinez-Montero[1], Maria Teresa Gonzalez Arnao[2] and
Florent Engelmann[3,4]
[1]University of Ciego de Avila/Bioplantas Centre
[2]Universidad Veracruzana,
[3]IRD, UMR DIAPC
[4]Bioversity International
[1]Cuba
[2]Mexico
[3]France
[4]Italy

1. Introduction

Sugarcane (*Saccharum* sp. hybrids) is a crop of major importance, which is cultivated on a large scale in tropical and subtropical regions primarily for its high sucrose content. Cultivated pineapple (*Ananas comosus* (L.) Merrill, which is now called Ananas comosus var comosus) belongs to the family Bromeliaceae. It is economically the fourth most important crop worldwide in terms of tropical fruit production and follows banana, mangoes and citrus. One of the main drawbacks faced by sugarcane and pineapple agriculture worldwide is the vegetative (i.e. asexual) nature of its conventional propagation. The consequence is that plants in the field must be replaced at intervals ranging from 1 to 5 years, a process that is costly, tedious and time-consuming. Furthermore, if the planting material is of low quality, yields decrease and more tillage is needed. The crops are exposed to natural disasters, while the propagation system leads to systemic disease transmission, and natural selection and plagues also take their toll. Moreover, the industry is in dramatic need of planting material, which cannot be produced in sufficient quantities to meet the demand using classical macropropagation techniques.

In vitro culture techniques have been extensively developed and applied for several thousand plant species including sugarcane and pineapple. Their uses are of high interest for multiplication, conservation and transformation of plant germplasm. Indeed, they allow the multiplication of plant material with high multiplication rates in an aseptic environment, reduction of space requirements, genetic erosion is reduced under optimal storage conditions, and minimized of the expenses in labour costs. Moreover, tissue culture systems greatly facilitate the international exchange of germplasm as the size of the samples is drastically diminished and they can be shipped in sterile conditions. Different *in vitro* conservation methods are employed, depending on the storage duration requested. For short- and medium-term storage, the aim is to reduce growth and to increase the intervals between subcultures. For long-term storage, cryopreservation, i.e. storage at ultra-low temperature, usually that of liquid nitrogen (−196°C), is the only current method ensuring long-term storage of germplasm from vegetatively propagated species. At this temperature, all cellular divisions and metabolic processes are stopped; therefore, plant material can thus be maintained without alteration or modification. Moreover, cultures are stored in a small volume, protected from contamination, requiring very limited maintenance.

This Chapter comprises two main sections focusing on the establishment, optimization and application of cryopreservation techniques to different tissues of *in vitro* sugarcane and pineapple cultures. The first part presents the cryopreservation protocols developed for sugarcane apices isolated from *in vitro* grown plants, embryogenic calluses and somatic embryos, as well as some analytical techniques (electrolyte leakage, protein content and lipid peroxidation products), used to describe the impact of the successive steps of the protocol on the physiological state of the cultures, which are also useful to refine the cryopreservation protocol. The effect of cryopreservation on the phenotypical development, both *in vitro* and in the field, of sugarcane plants regenerated material will be also presented. The second section presents the studies performed to set up and refine a cryopreservation protocol for apices of pineapple *in vitro* plantlets. The protocol established following the vitrification approach was successfully applied for the first time to shoot tips of three pineapple varieties, and then extended to nine pineapple accessions belonging to the *in vitro* collection of Bioplantas Centre in Cuba. In addition, we present the preliminary assays developed using callus of two pineapple cultivars. In the conclusion, we discuss the possibilities and prospects of utilisation of cryopreservation techniques for the long-term storage of other vegetatively propagated tropical plant species.

2. Cryopreservation protocols for sugarcane

Several review papers have been published, which provide lists of species which have been successfully cryopreserved (Cyr, 2000; Engelmann, 1997; Engelmann & Takagi, 2000; Sakai et al., 2002). For vegetatively propagated species, cryopreservation has a wide applicability in terms of species coverage, since protocols have been successfully established for roots and tubers, fruit trees, ornamentals, forestry species and plantation crops from both temperate and tropical origin (Engelmann, 2004; Kaczmarczyk et al., 2008; Engelmann, 2010; Engelmann, 2011).

In the case of sugarcane, cryopreservation protocols have been developed for various materials: apices of *in vitro* plantlets using the encapsulation-dehydration technique (Gonzalez-Arnao et al., 1993; Paulet et al., 1993); cell suspensions (Finkle & Ulrich, 1979; Gnanapragasam & Vasil, 1990) and embryogenic callus using classical freezing protocols (Eksomtramage et al., 1992; Gnanapragasam & Vasil, 1992; Jian et al., 1987) and simplified cryopreservation protocols (Martinez-Montero et al., 1998). Recently, it was published the cryopreservation procedure based on vitrification techniques for somatic embryos (Martinez-Montero et al., 2008).

2.1 Cell suspensions

The first experimental research on sugarcane cell suspension was accomplished at the end of 1970 and at the beginning of 1980 decades (Finkle & Ulrich, 1979; Finkle & Ulrich, 1982; Ulrich et al., 1979; Ulrich et al., 1984). These authors demonstrated that resistance of cells to freezing to -23°C and -40°C was possible with little decrease in survival by using mixtures of glucose, dimethylsulfoxide and polyethylene glycol as cryoprotectants. However, no plants were recovered from the cryopreserved cell suspension.

Later on, Gnanapragasam & Vasil (1990) reported that efficient plant regeneration was obtained from a cryopreserved embryogenic cell suspension of one commercial sugarcane hybrid established from leaf derived callus. They observed pregrowing the cells for three days in Murashige & Skoog (1962) basal medium supplemented with 0.33 M sorbitol was essential to the process. A regeneration efficiency of 92% was obtained and plants regenerated from cryopreserved cells, and grown to maturity in the greenhouse, were morphologically identical to regenerated control plants. Later, there were not detected differences at molecular level using RFLP technique comparing plants regenerated from cryopreserved and control cells for three sugarcane hybrids (Chowdhury & Vasil, 1993).

2.2 Embryogenic callus

The first success for cryopreservation of sugarcane embryogenic callus was obtained by Ulrich et al., (1979) for the hybrid H50-7209. It was a pretreatment using a combination of 10% polyethylene glycol, 8% glucose and 10% DMSO, freezing rate of 2°C.min^{-1} until a first transfer temperature of -40°C and freezing rate of 5°C.min^{-1} until second transfer temperature of -80°C. However, the recovery of cryopreserved callus was achieved only with root regeneration. Ulrich et al., (1984) obtained after modifications of the same protocol a limited number of albino plantlets from cryopreserved calluses.

Later on, high survival rates (ca. 90%) and recovery of whole plants were obtained by Jian et al., (1987), Eksomtramage et al., (1992) and Gnanapragasam & Vasil (1992). The conditions defined were different from that used by Ulrich et al., (1979, 1984). For cryoprotective treatment, a mixture of sorbitol and DMSO was used by Jian et al., (1987) and Gnanapragasam & Vasil (1992); Eksomtramage et al., (1992) employed a mixture of sucrose and DMSO. Freezing conditions were also different: 1°C.min^{-1} from 0°C to -10°C, and kept for 15min at the same freezing rate from -10°C down to -40°C and kept for 1-5 h, and finally immersed into liquid nitrogen (Jian et al., 1987); or 0.5°C.min^{-1} down to -40°C or -45°C with no plateau at the end of the controlled freezing sequence (Eksomtramage et al., 1992 and Gnanapragasam & Vasil, 1992). Moreover, the technique developed by Eksomtramage et al., (1992) was successfully applied to calluses of 10 varieties.

These authors have followed the strategy known as dehydration by extracellular freezing, which uses a controlled freezing regime (Withers & King, 1980). However, this procedure requires expensive and sometimes complex programmable freezing devices, limiting its use to laboratories specializing in cryopreservation (Ashmore, 1997; Reed, 2001). Furthermore, their research has been focused on the cryopreservation of sugarcane calli obtained from segments of immature leaves belonging to *in vitro* cultured plants; however, such explants are known to have a limited morphogenetic capacity (Krishnaraj & Vasil, 1995) and it is widely acknowledged that immature embryos, as well as young inflorescences, are physiologically better explants for calli production because they retain their embryogenic capacities (Merkle et al., 1995).

2.2.1 Optimization of methodology for sugarcane callus

Our research team (Martinez-Montero et al., 1998, 2006), using the cryo-research for sugarcane callus described above as starting point, published the results for establishing step by step a methodology for the cryopreservation of sugarcane calli with embryogenic structures obtained from immature inflorescence (Figure 1). We optimized the following aspects according to the *in vitro* survival and regeneration (plants per 500 mg of calli) percentages for: Selection of the cooling procedure, the effect of the cooling procedure and of the type of alcohol, the effect of the induction time of extracellular ice crystals, the effect of post-subculture time, the effect of sucrose and dimethylsulfoxide concentration in the cryoprotective medium, and the effect of the pre-freezing time.

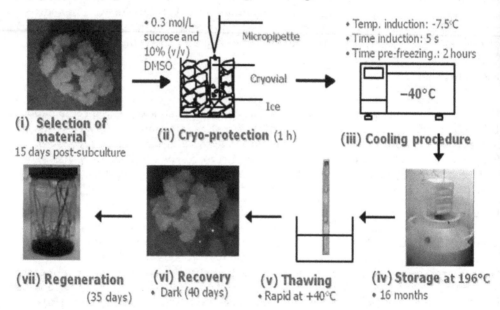

Fig. 1. Optimized methodology for the cryopreservation of sugarcane calli with embryogenic structures.

Firstly, we based on the results carried out by Maddox et al. (1983) and Withers (1985) who successfully used uncomplicated freezing procedures for cellular suspensions of *Nicotiana* and *Musa*, respectively (Figure 2). We evaluated the application of one of these devices as an alternative to establish techniques for the cryopreservation of sugarcane calli with embryogenic structures (i.e., cooling rate controlled by a computer-coupled programmable freezer).

As results was detected a survival after storage in liquid nitrogen for both cooling procedures implying the existence of a protective dehydration process that allows the vitrification of some cells without the formation of intracellular ice crystals. However, from

Cryopreservation of Tropical Plant Germplasm with Vegetative Propagation – Review of Sugarcane (Saccharum spp.) and Pineapple (Ananas comusus (L.) Merrill) Cases

127

a practical point of view, seeding ice crystals is much more difficult when using the procedure proposed by Maddox et al., (1983).

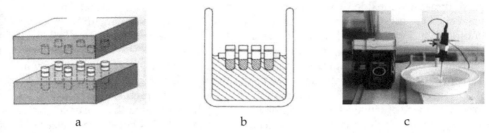

a   b   c

Fig. 2. Devices used by Maddox et al. (1983) (A), Withers (1985) (B) and our research group (C) for the establishment of cryopreservation procedures.

In general, the "classical" cryopreservation protocols provide insufficient detail on seeding ice crystals step (Martinez-Montero et al., 2006). For instance, although the analysis of this parameter must have been unavoidable for the development of the protocols of Jian et al., (1987) and Gnanapragasam & Vasil (1992), these analyses are not described in their articles; and even Eksomtramage et al., (1992), who first mentioned the need for this step when freeze sugarcane calli, provide little detail on its implementation.

Optimizing the effect of post-subculture time on the survival and regeneration of plants from cryopreserved calli the results reported by Jian et al., (1987) and Martinez-Montero et al., (2006) coincided. We founded that survival after cryopreservation is associated with the selection step during the post-subculture period, reaching a maximum at 15 days post-subculture. Moreover, we correlated this finding to the physiological state of the calli before cryopreservation and measured the growth of the calli. These results are the basis for a rational selection of the material to be cryopreserved, since several authors have shown that there is a correlation for different species between the phase of active growth of the calli and its performance upon cryopreservation (Reinhoud et al., 2000; Withers, 1985; Yoshida et al., 1993). It has been proven that the morphology of the cultured cells has a marked influence on cryotolerance. In most species, only small cells with a highly dense cytoplasm, usually found in small cellular aggregates in the periphery of the callus, survive after cryopreservation; whereas large, vacuolated cells are damaged during freezing (Kristensen et al., 1994; Withers, 1985).

We also founded the decrease in survival and regeneration when using sucrose concentrations higher than 0.3 M. The importance of sucrose tolerance within this setting is determined by the role of this disaccharide in the regulation of the hydric potential of the cells (Tetteroo, 1996); sucrose has also occasionally been considered an inducer of cellular division and differentiation (Feher, 2003). Furthermore, there is evidence suggesting that sucrose functions as a genetic regulatory signal for genes coding for enzymes and proteins involved in transport and storage (Lunn & MacRae, 2003). Additionally, Ausborn et al., (1994) and Turner et al., (2001) detected that sucrose stabilizes the lipid bilayers on the membranes by forming disaccharide-lipid hydrogen bonds, whereas Niu et al., (1997) founded that the right amount of intracellular sucrose can protect a number of enzymes from ion-mediated toxicity.

According to our results, the sugarcane calli did not survive the cryopreservation procedure when dimethylsulfoxide was omitted from the cryoprotective mixture; and both the survival and plant regeneration percentages rose steadily with increasing dimethylsulfoxide concentrations, up to 10%. However, in clear contrast with the results obtained when testing different amounts of sucrose, there are no differences in survival between the cryopreserved and non-cryopreserved samples at concentrations higher than the optimum (10% v/v in this case), and the contrast is even starker when comparing plant regeneration rates, where the cryopreserved material performs even better than the non-cryopreserved calli. These results agree with those of Finkle et al., (1985) for rice cells, who concluded that the effects achieved by using dimethylsulfoxide are paradoxical, since although this substance is clearly toxic, but inhibits the growth of ice crystals during cryopreservation.

Our data obtained during the experiment for dimethylsulfoxide tolerance are coherent with the reports for other biological systems, associated to the high degree of toxicity of dimethylsulfoxide (Arakawa et al., 1990; Fahy et al., 1990). Kartha et al., (1988) detected that dimethylsulfoxide produces an inhibition of 35 to 42% on the growth of embryogenic cultures of white spruce when used at a concentration higher than 5% (v/v), and Klimaszewska et al., (1992) reported a 28% reduction in the growth of embryogenic tissues from black spruce when treated with 15% (v/v) dimethylsulfoxide; this effect, according to the microscopic observations of these authors, is due to the induction by this substance of a strong plasmolytic effect at the cellular level.

However, and in spite of these findings, dimethylsulfoxide has been, and still is used as a cryoprotectant during storage at ultra-low temperatures. According to Engelmann, (2000) this apparent paradox is due to the fact that dimethylsulfoxide is always used as part of a cryoprotective mixture, rather than individually. Arakawa et al., (19990) have provided evidence that the toxicity of dimethylsulfoxide in isolated proteins is mediated by hydrophobic interactions, which are favored at increasing temperatures; in this context, this effect is minimized by the use, during preculture, of sucrose at 0°C, which induces the biosynthesis of proteins that neutralize the toxic effects of this agent (caused by its interaction mainly with lysine residues) (Anchordoguy et al., 1991; Klimazewska et al., 1990; Swan et al., 1999).

Although the exact cryoprotective mechanism of dimethylsulfoxide at ultra-low temperatures remains unknown, it is widely acknowledged that it depends on the colligative properties of this penetrating compound; that is, dimethylsulfoxide affects the formation of ice crystals by decreasing the equilibrium freezing point of the solution, in direct dependence on its molar concentration (Kinoshita et al., 2001; McGann, L.E. & Walterson, 1987). Dimethylsulfoxide, as a cell-penetrating agent, also decreases the intracellular concentration of toxic electrolytes on unfrozen cells (Finkle et al., 1985).

Anchordoguy et al., (1991) suggested, furthermore, that there is another, not colligative mechanism for dimethylsulfoxide-mediated cryoprotection, which involves ionic interactions between the oxygen atom from this molecule and phospholipid bilayers. Such a mechanism would stabilize the cell membranes during the freeze-thaw cycle. The findings by us related wit better plant regeneration percentages from cryopreserved calli as compared to calli which had not been cryopreserved could be similar to those of Aronen et al., (1999) in embryogenic cultures of *Abies cephalonica*. We later founded that the storage in liquid nitrogen eliminates a high proportion of cells which had been previously damaged by dimethylsulfoxide, since only small, meristematic cells survive this treatment. Furthermore, the use of cryoprotective mixtures containing other agents greatly minimizes the inherent toxicity of dimethylsulfoxide.

On the other hand, it is recognized that the process of apoptosis (programmed cell death) is not circumscribed to animals, but also occurs in plants, where it is used for the selective elimination and suicide of unwanted cells (Krishnamurthy et al., 2000). According to Joyce et al., (2003), among the cells undergoing this process are those which have sustained high levels of *in vitro* stress, which can compromise their physiology. Such a mechanism might, therefore, be involved in dimethylsulfoxide-mediated toxicity.

After using the simple freezing procedure proposed by Martinez-Montero et al. (1998), it was determined that, apparently, the best dehydration levels are reached by the sugarcane calli when kept for 2 or 3 hours at -40 °C. It should be noted that the survival percentages achieved in this study were comparable to the best values obtained by Jian et al., (1987) and Eksomtramage et al., (1992). Survival rates did not increase with longer pre-freezing times, probably due to excessive dehydration of the material. Studies based on the use of nuclear magnetic resonance spectroscopy in *Catharanthus roseus* cells (Chen et al., 1984), vegetative apple buds (Tyler et al., 1988) and different tissues from *Rhododendron japonicum* (Ishikawa et al., 2000) have determined that the optimum pre-freezing time for a specified pre-freezing temperature depends on the amount of water still remaining inside the cells. Tyler et al., (1988) proved the need for the pre-freezing step when they showed that the incubation of samples at an intermediate negative temperature before immersion in liquid nitrogen would result in a better performance of the cryopreserved material after thawing.

Finally, the optimized protocol carried out by our team took into account the *in vitro* survival and regeneration (plants per 500 mg of sugarcane calli) percentages and was validated for: a) three varieties (CP52-43, C1051-73, C91-301) (Table 1) ; b) explants obtained either from immature inflorescences or immature leaves from *in vitro* plants; c) calli stored for up to 16 months under liquid nitrogen, belonging to the CP52-43 variety (Table 2).

Variety		Survival (%)	Regeneration (plants per 500mg of calluses)
-LN	CP52-43	98,8 a	230 a
	C91-301	69,5 c	72 c
	C1051-73	44,1 d	55 d
+LN	CP52-43	89,0 b	150 b
	C91-301	38,8 d	42 e
	C1051-73	22,2 e	25 f
Typical Error		0,190	1,421

Table 1. Effects of optimized cryopreservation protocol on survival and plant regeneration produced from control (-LN) and cryopreserved (+LN) sugarcane embryogenic calluses (varieties CP52-43, C91-301 and C1051-73). *Means within columns followed by the same letter are not significantly different (ANOVA p < 0,05 Tukey,). Data were transformed for statistical analysis in accordance with x'= 2 arcsine ((x/100)^{0,5}) and with x'= (0,5 + x)^{0,5} for percentage of survival and plant regeneration, respectively.*

	Time (months)	Survival (%)	Calluses that regenerated plants (%)	Regeneration (plants per 500mg of calluses)
-NL	1	100 a	100,0 a	225,2 a
	4	98,6 a	100,0 a	224,3 a
	8	96,7 ab	66,7 b	77,0 c
	12	97,5 ab	7,1 c	19,1 d
	16	11,5 c	0,0 d	0,0 e
+NL	1	90,6 b	96,7 a	149,3 b
	4	87,6 b	97,8 a	142,9 b
	8	88,0 b	93,3 a	135,1 b
	12	86,4 b	97,0 a	141,7 b
	16	90,0 b	97,5 a	140,3 b
Typical Error		0,189	0,253	1,312

Table 2. Effect of extended storage duration on the survival and plantlet produced from control (-LN) and cryopreserved (+LN) sugarcane embryogenic calluses (variety CP52-43). *Means within columns followed by the same letter are not significantly different (ANOVA $p < 0,05$ Tukey,). Data were transformed for statistical analysis in accordance with $x' = 2$ arcsine $((x/100)^{0,5})$ and with $x' = (0,5 + x)^{0,5}$ for percentage of survival and plant regeneration, respectively.*

2.3 Apices

The first attempt to freeze sugarcane apices were carried out by the group of Bajaj et al., (1987) using apices from *in vivo* plants. This material was pretreated with a mixture of 5% (v/v) of each DMSO, sucrose and glycerol during 45min. Then the freezing was accomplished by rapid immersion in liquid nitrogen of samples. However, the apices recovery was very scarce and with only small callus formation without plant regeneration.

Later on, research for the development of a cryopreservation protocol for sugarcane apices was carried out in the framework of collaborative program involving IRD (Institut de recherche pour le développement, Montpellier, France), CIRAD (Centre de coopération internationale en recherche agronomique pour le développement, Montpellier, France), CNIC (Centro Nacional de Investigaciones Cientificas, Havana, Cuba), IPGRI (International Plant Genetic Resources Institute, Rome, Italy) and FAO (Food and Agriculture Organisation of the United Nations, Rome, Italy).

A cryopreservation process using encapsulation/dehydration was set up for apices sampled on *in vitro* plantlets of sugarcane by Paulet et al., (1993) in CIRAD, Montpellier, France. After dissection, apices were cultured for one day on standard medium and then encapsulated in medium with 3% alginate. Optimal conditions comprised preculture for 2d in liquid medium with 0.75M sucrose, desiccation for 6h under the laminar flow or for 10–11 hours with silicagel followed by rapid freezing and slow thawing. Survival after freezing in liquid nitrogen ranged between 38 and 91% for the 5 varieties experimented.

Later on, Gonzalez-Arnao et al. (1993) in CNIC, Havana Cuba investigated the effect of sucrose concentration during the pregrowth treatment and of freezing procedure on the survival of encapsulated apices of six sugarcane varieties. The optimal sucrose concentration was 0.75 M during 24h. We showed that encapsulated apices of sugarcane could withstand freezing in liquid nitrogen using various freezing procedures. Growth recovery of apices after thawing was very rapid and direct, due to the fact that most cells of the apical region had been only slightly harmed, as revealed by histological examination.

Moreover, our group studied apices sampled on *in vitro* plantlets of different varieties and could be cryopreserved using the encapsulation-dehydration technique and stored for one year at the temperature of liquid nitrogen without modification in their recovery percentage (Gonzalez-Arnao et al., 1999). By contrast, apices placed at -70°C or -25°C lost viability very rapidly. There are several explanations for this result: even though vitrification of internal solutes has been observed during freezing of these materials, including sugarcane devitrification and recrystallization processes, which are detrimental to 'cellular integrity, take place at these temperatures (Gonzalez-Arnao et al., 1996). These contrasting results might be linked to the presence of higher levels of residual free water in the latter systems, which would recrystallize rapidly at these temperatures and result in the death of the explants. At lower temperatures comprised between -135 and -196"C, no differences were noted in the regrowth capacity of all materials mentioned above whatever the storage duration tested.

It is interesting to note that, even though the two protocols set up were slightly different, the average results obtained on a total of 15 sugarcane varieties (8 frozen with the CNIC protocol, 7 with the CIRAD protocol) were similar (Table 3). It should also be noted that different varieties showed different sensitivities to preculture and desiccation, and to preculture, desiccation and freezing. However, there was only one case (Ja 60-5) where the difference between control and cryopreserved samples was very high, 70% and 24% survival, respectively. Both protocols are thus potentially applicable to a large range of varieties without any need for further adaptation.

Recently, *in vitro* shoot tips of two clones were successfully cryopreserved using encapsulation-dehydration according to Gonzalez-Arnao et al., (1993) and droplet-vitrification with two vitrification solutions, PVS2 and PVS3 (Barraco et al., 2011). For both clones, encapsulation-dehydration induced significantly higher recovery, reaching 60% for clone H70-144 and 53% for clone CP68-1026, compared with droplet-vitrification in which recovery was 33-37% for clone H70-144 and 20-27% for clone CP68-1026. Optimal conditions included preculture of encapsulated shoot apices for 24 h in liquid medium with 0.75 M sucrose and dehydration with silica gel to 20% moisture content (fresh weight basis) before direct immersion in liquid nitrogen. With both protocols employed, regrowth of cryopreserved samples, as followed by visual observation, was always rapid and direct.

Variety	Survival	
	-LN	+LN
C 87-51 (*)	90	70
C 266-70 (*)	90	86
B 34104 (*)	60	67
B 4362 (*)	88	74
Ja 60-5 (*)	70	24
IAC 5448 (*)	50	38
POJ 2878 (*)	56	60
CP 70-1133 (*)	80	60
CP 68-1026 (**)	100	64
B 69566 (**)	100	91
Co 6415 (**)	80	64
Co 740 (**)	80	38
IAC 5118 (**)	50	38
My 5514 (**)	83	75
Q 90 (**)	100	82

Table 3. Survival of control (-LN) and cryopreserved (+LN) apices using the encapsulationdehydration technique according to the protocol described by Gonzalez-Arnao et al. (1993, *) and Paulet et al. (1993, **).

2.4 Somatic embryos

In Cuba, a micropropagation protocol based on the artificial seed technology has been established for sugarcane, which uses somatic embryos produced on semi-solid medium, an alternative which allows mass multiplication of plants from elite varieties (Nieves et al., 2001; Tapia et al., 1999). However, this protocol has an important limiting factor, which lies with the necessity of safely storing somatic embryos for the long-term (Benson, 2008). Establishing a cryopreservation protocol for somatic embryos would allow solving this problem.

In this sense our group (Martinez-Montero et al., 2008) compared three vitrification-based cryopreservation techniques, viz. vitrification, encapsulation-vitrification and droplet-vitrification for cryopreserving sugarcane somatic embryos. No viability was achieved using the vitrification procedure. The comparison of the recovery pattern of cryopreserved somatic embryos showed that droplet-vitrification procedure was more efficient than encapsulation-vitrification (Table 1) based on the presence of green colour in somatic embryos and on the percentage of clumps with embryos converted into plants. Moreover, the presence of callus together with converted plants was observed with the encapsulation-vitrification procedure.

Protocol	Presence of green colour in SE (%)	Clumps with SE converted into plants (%)	Presence of callus together with converted plants (%)
Encapsulation-vitrification *	8 b	5 b	67 a
Droplet-vitrification**	25 a	18 a	0 b
SEM	0.22	0.21	0.31

Means with different letters are statistically different (t-test, $P < 0.05$). The data were transformed before the analysis using $x' = 2 \arcsin((x/100)^{0.5})$. (Number of samples = 144).
*. The selected samples were encapsulated with 3 % (w/v) Na-alginate and loading solution (2 M glycerol + 0.4 M sucrose). The beads formed were dehydrated at 0 °C for 10 min in PVS2 before direct immersion in liquid nitrogen.
** The selected samples were loaded in 2 M glycerol + 0.4 M sucrose for 20 min at 25 °C. Then, they were placed on the filter paper, transferred in cryotubes and dehydrated at 0 °C for 80 min with PVS2. The samples in cryotubes were transferred in droplets of PVS2 solution (10 µl) placed on aluminium foil strips and the strips were immersed in liquid nitrogen.

Table 4. Effect of the cryopreservation protocol on the recovery pattern of sugarcane somatic embryos (SE).

Untreated embryos were white (Fig. 3A & B). Cryopreserved embryos were white to yellow when they were placed on recovery medium; viable embryos turned yellow to green after about 2 weeks; they converted to plants within an additional 2 week period and produced green shoots and roots (Fig. 3 C & D). Callus formation was not observed in germinated embryos and no secondary embryos were produced after the droplet-vitrification procedure (Fig. 3 C). However, callus appeared together with germinated embryos after encapsulation-vitrification (Fig. 3D).

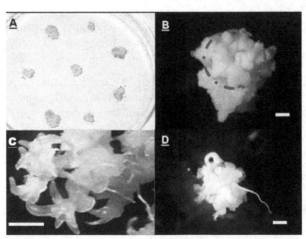

Fig. 3. Initial embryogenic sugarcane callus (A); clumps of somatic embryos selected for cryopreservation experiments (dashed line) (B); recovered clumps of somatic embryos after cryopreservation and 4 weeks after transfer to MS medium under light conditions (C, using droplet-vitrification procedure; D, using encapsulation-vitrification procedure) (bar = 1mm).

The obtained results by Martinez-Montero et al., (2008) contrasted with what is generally observed in the literature, as vitrification is the most frequently employed vitrification-based procedure and it has been applied to a large number of species (Panis & Lambardi, 2006; Sakai & Engelmann, 2007). However, the number of successful reports of application of the droplet-freezing and encapsulation-vitrification techniques is increasing steadily (Engelmann, 2011).

Sugarcane somatic embryos proved very sensitive to PVS2, even though the PVS2 treatment was performed at 0°C, which usually reduces the toxicity of the cryoprotectant solution (Benson, 2008). This high sensitivity rendered the utilization of the vitrification procedure impossible and alternative options had to be sought.

One of the options tested for cryopreservation of sugarcane somatic embryos was the encapsulation-vitrification technique, as developed by Matsumoto et al., (1995). These authors suggested that the toxicity of the PVS2 solution could be reduced by encapsulating the explants in alginate beads. Encapsulation also made the manipulation of the material easier. The positive effect of employing this technique was confirmed by the results, as some viability was achieved after cryopreserving sugarcane embryos using encapsulation-vitrification.

We also tested the droplet-vitrification technique with sugarcane embryos (Martinez-Montero et al., 2008). Droplet vitrification combines the procedure called droplet-freezing, which has been established with cassava (Kartha & Engelmann, 1994) and applied notably to potato (Schäfer-Menuhr et al., 1997) and asparagus shoot tips (Mix-Wagner et al., 2000), in which explants are cooled in a droplet of cryoprotectant solution with the vitrification procedure (Sakai et al., 1990), since explants are cooled in a droplet of PVS2 solution. Droplet-vitrification is relatively easy to implement and generally ensures high recovery after cooling (Sakai & Engelmann, 2007). One of the advantages of this technique is the high cooling and warming rates achieved, compared with others procedures (Benson, 2007; Panis et al., 2005). These high cooling/warming rates ensure complete vitrification during cooling and reduce the risks of devitrification during warming of samples, which is important to avoid the lethal effects of intracellular ice crystal formation (Benson, 2008).

Moreover, Volk & Walters (2006) concluded that PVS2 imparts its effect in the previtrified solution, and at lower temperature the cryoprotectant restricts the mobility of water molecules, so that they are unable to nucleate and ice crystals are not allowed to growth. Benson (2008) empathized that cryoprotection using droplet-vitrification involves a somewhat different principle, due to the behavior of water molecules contained in micro-droplets of vitrification solution. If the biophysical conditions are optimal the droplets can become vitrified on direct exposure to liquid nitrogen.

3. Use of analytical techniques for sugarcane cryopreservation protocols

In the past, cryopreservation protocols have generally been developed using an empirical approach. However, considerable advances have been made in recent years in the use of analytical tools to enhance our current knowledge of the damages induced in biological tissues by cryopreservation (Engelmann, 2011). Various biophysical, biochemical and histo-cytological techniques are available for this purpose (Harding, 1999). Such analytical tools

Cryopreservation of Tropical Plant Germplasm with Vegetative Propagation – Review of Sugarcane
(Saccharum spp.) and Pineapple (Ananas comusus (L.) Merrill) Cases

135

allow the detection of those components of a cryopreservation method which cause the most damage. Usually, these studies are correlated with survival responses and viability testing. However, the application of analytical tools for plant cryopreservation studies is still very scarce and in some cases they are costly to implement and complex to evaluate (Verleysen et al., 2004). Apart mention need the excellent review by Benson (2008) in which it is exposed that contemporary cryopreservation research is now supported by advanced biomolecular or 'omics' technologies, creating a new knowledge base which will hopefully help to solve some of the more difficult cryobiological challenges. However, it will become increasingly so as stakeholders invest in areas commonly interested in low temperature research. Therefore, our research experience is only limited to use non costly and complex analytical techniques yet.

3.1 Effect of cryopreservation on the structural and functional integrity of cell membranes of sugarcane embryogenic callus

Cell membranes are one of the main targets of numerous stressing events, including cryopreservation (Benson, 2007; Fahy et al., 1984; Engelmann, 2011). Various markers, including electrolyte efflux, lipid peroxidation products and cell membrane protein content, reflect the structural and functional integrity status of cell membranes after exposure to such stressing events (Harding, 1999; Verleysen et al., 2004).

Measurement of electrolyte leakage has been used notably for studying the desiccation and cryopreservation sensitivity of various recalcitrant seed species (Sun, 1999). Lipid peroxidation profiles have been used as markers of cell membrane damage during freezing of rice cell suspensions and of the coenocytic alga *Vaucheria sessilis* (Benson et al., 1992; Fleck et al., 1999). Watanabe et al., (1999) have shown that the acquisition of tolerance to cryopreservation of rice cells was related to changes in protein metabolism. An increasing number of proteins and peptides that might contribute to freezing tolerance by reducing the effects of dehydration associated with freezing have been identified (Thomashow, 1999). In the same way, Thierry et al., (1999) have observed in carrot somatic embryos the over-accumulation of boiling-stable proteins, which seems to be related to an increase in tolerance to cryopreservation. Besides, some enzymes, which are induced by low temperature, such as fatty acid desaturase and sucrose phosphate synthase, also contribute to freezing tolerance (Guy, 1999).

Our research group studied the effect of cryopreservation on the structural and functional integrity of cell membranes of sugarcane embryogenic calluses by measuring electrolyte leakage, lipid peroxidation products and membrane proteins (Fig. 4). Firstly, we showed (Martinez-Montero et al., 2002a) that survival and plantlet production were lower with cryopreserved sugarcane embryogenic calluses in comparison with unfrozen control calluses. However, the differences observed between control and cryopreserved calluses in the parameters studied to evaluate membrane structural and functional integrity, including electrolyte leakage, total cell membrane protein content, malondialdehyde and other aldehyde content were only transitory. Indeed, they had all disappeared within 3-4 days after freezing.

Electrolyte leakage, measured to evaluate the overall effect of cryopreservation on the semi-permeability of plasma membranes, revealed a partial loss of membrane semipermeability in callus cells. The transitory character of the electrolyte efflux observed indicates that no dramatic mechanical cell membrane injuries were caused by cryopreservation; rather only reversible lesions were induced by this treatment. As part of this dynamic process, the electrolytes released by damaged cells may have been taken up by living cells.

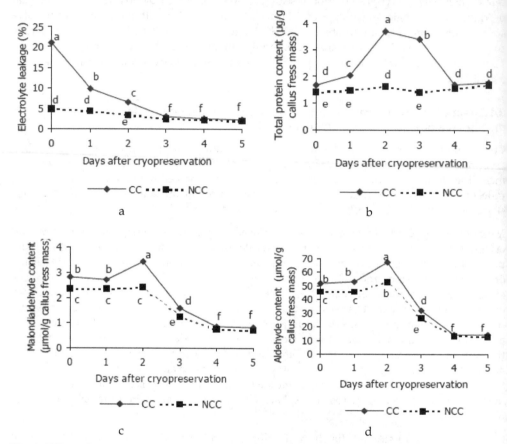

Fig. 4. Effect of cryopreservation on the structural and functional integrity of cell membranes of sugarcane embryogenic calluses by measuring electrolyte leakage (A), total proteins (B), malondialdehyde content (C) and other aldehyde content (D). *Data followed by the same letter are not statistically different (ANOVA, Duncan, p<0.05). CC: cryopreserved callus. NCC: non-cryopreserved callus.*

Freezing injury induces the production of free radicals, mainly reactive oxygen species (ROS) (Benson et al., 1992). The ROS signaling pathway is mainly controlled by the production of, and balance between, pro- and antioxidants and the perturbation of ROS homeostasis (Mittler et al., 2004). These changes are perceived by various proteins, enzymes and receptors which influence different developmental, metabolic and defense pathways. Free radicals then attack the lipid fraction of membranes, resulting in the formation of unstable lipid peroxides. These compounds breakdown to form toxic secondary oxidation products (Esterbauer et al., 1988) such as aldehydes, including malondialdehyde and other aldehydic products.

According to our results the main factors affecting sugarcane callus cell membrane damages and electrolyte efflux might thus be oxygen reactive species instead of malondialdehyde and aldehydes themselves, since the highest concentration of these compounds was reached later than the highest level of electrolyte leakage. However, it is also possible that the damages noted after cryopreservation could have been caused by the loss of cellular integrity due to the formation of ice crystals and to the cryoprotectants employed, which could damage the membranes.

We also showed that the content in malondialdehyde and other aldehydes in the microsomal fraction were higher for cryopreserved calluses than unfrozen controls, but only during the first three days after cryopreservation. Benson et al., (1992) have obtained similar results for malondialdehyde with cryopreserved rice cell suspensions. Therefore, they suggested that freezing stress could have caused disruption and uncoupling in some metabolic pathways as reported by Fleck et al. (1999) and Dumet et al., (2000) with other biological systems. This could have led to the production of free radicals, thus promoting lipid peroxidation in the cellular membranes of calluses at a very early post-thaw recovery stage.

Variations were also observed in control calluses, concerning mainly electrolyte leakage and lipid peroxidation. The significantly increased levels of malondialdehyde and aldehydes measured during the first 3 days in control calluses might be a result of mechanical membrane damage caused by cutting when preparing the starting material. Fleck et al., (1999) described an increase in lipid peroxidation products after cutting algae filaments into sections. In addition, transfer of material to fresh medium itself is another stress source that may cause an increase in malondialdehyde and aldehydes (Benson, 2007).

The concentration of lipid peroxidation products decreased from the second day onwards and reached a constant value on the fourth day in both frozen and control calluses (Martinez-Montero et al., 2002a). This decrease must have been caused by the activation of antioxidant defense mechanisms. Plants produce antioxidant molecules and have scavenging systems (ß-carotenes, tocopherol isomers, ascorbic acid, glutathione) and enzymatic free radical processing systems (superoxide dismutase, catalase, glutathione reductase, ascorbate peroxidase and various other enzymes) as a protective response to stresses (Leprince et al., 1993). Those antioxidant systems are directly activated by oxidative stress and, consequently, diminish the levels of ROS and thiobarbituric acid reactive substances in cells. Martinez-Montero et al., (2002a) suggested additional experiments to be performed to measure the concentration of such antioxidant molecules and the activity level of the above-cited enzymes in sugarcane embryogenic calluses in relation to cryopreservation.

An increase in cell membrane-related proteins has been described as a response to dehydration and freezing stress (Ausborn et al., 1994). Such proteins are produced as a protective mechanism to preserve membrane structure, ion sequestration and chaperon-like functions (Thierry et al., 1999). According to Martinez-Montero et al., (2002a) the total microsomal fraction protein content was higher in cryopreserved sugarcane calluses during the first 3 d after thawing. This increase was concomitant with an increase in malondialdehyde and aldehyde concentration. They hypothesized that some of the proteins induced by the freeze-thaw cycle may play a role in decreasing the malondialdehyde and aldehyde levels, in addition to the other functions mentioned above.

Even though electrolyte leakage, malondialdehyde, aldehyde and cell membrane protein contents became similar in control and cryopreserved samples 4 d after cryopreservation, cryopreservation consistently affected callus survival and plantlet regeneration (Martinez-Montero et al., 2002a). However, lipid peroxidation products (such as malondialdehyde and aldehydes) might have impaired various cell functions in the sugarcane embryogenic calluses by cross-linking to macromolecules such as DNA and proteins to form mutagenic compounds as reported by Yang & Scaich, (1996) for animal cells. Moreover, the free radicals induced by freezing stress are considered both cytotoxic and genotoxic because they are capable to modify protein structure, to form complexes with DNA and enzymes and to inhibit nucleic acid synthesis (Esterbauer et al., 1988; Grune et al., 1997, Spiteller, 1996). Such impairments might have affected the totipotency of these callus cells.

3.2 Use of electrolyte leakage technique for sugarcane somatic embryos

To allow a quick, reliable prognosis of the experiments performed and to refine the optimal conditions for cryopreservation of somatic embryos, viability was estimated using an electrolyte efflux test by our research group (Martinez-Montero et al., 2008). Firstly, four dissected clumps were either immediately incubated in 20 ml double-distilled water or immersed directly in liquid nitrogen before incubation. The percentage of living cells (cell viability) was calculated by making mixtures of cooled and non cooled clumps (ca. 20 mg total fresh weight). Conductivity of the water was measured before (C0) and after 5, 15 or 25 h of imbibition (Cx). Samples were then autoclaved (30 min at 112°C, 107 kPa) and cooled down to room temperature for 4 h to determine the total conductivity (Ctotal). The percentage of electrolyte leakage was calculated from the ratio: (Cx – C0)*100/(Ctotal – C0). Lastly, a regression analysis was made between the results of the electrolyte leakage test and the cell viability (both were expressed in percentages). The best model that was suitable to represent the experimental data constitutes a standardized curve for analysis of cell viability during all experimentations.

The effect of the imbibitions duration in double-distilled water on the leakage of electrolytes in mixtures of cooled and non-cooled clumps by regression analysis demonstrated that there was a significant linear relationship ($\alpha = 0.05$) between the electrolyte efflux and the viability of somatic embryos and high coefficients of determination ($R^2 > 0.88$) were obtained (Fig. 5). It was clearly observed that a greater functional relationship existed in the lineal equation for imbibitions periods of clumps between 15 and 25 h. Our results confirmed that an equilibration period was necessary for accurate measurement of cell leakage and the electrolyte leakage was practically complete (85% of the electrolytes leaked in a sample with 100% of cooled clumps) after 15 h only. Moreover, the values of electrolyte leakage indicated that close to 10% of the somatic embryos were damaged after dissection for the cryopreservation experiments. Therefore, our test proved useful and precise as it was not only good for distinguishing between living and dead tissues but also for quantifying small differences in the amount of viable tissues.

Cryopreservation of Tropical Plant Germplasm with Vegetative Propagation – Review of Sugarcane
(Saccharum spp.) and Pineapple (Ananas comusus (L.) Merrill) Cases

139

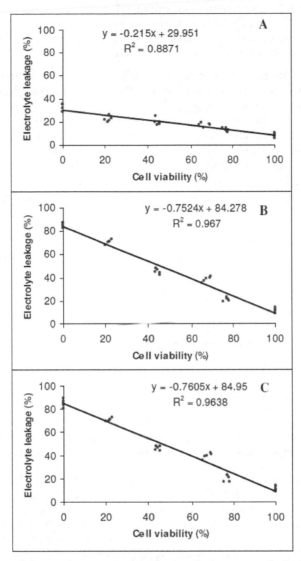

Fig. 5. Effect of the imbibition duration in double-distilled water on the leakage of electrolytes in mixtures of cooled and non-cooled clumps by regression analysis to determine the cell viability of sugarcane somatic embryos. A) after 5 h; B) after 15 h; C) after 25 h.

3.2.1 Optimization of methodology for somatic embryos

The loading treatment is an essential step to achieve high post-rewarming survival of cryopreserved sugarcane somatic embryos because it induces or enhances tolerance of samples to PVS2 treatment (Panis & Thinh, 2001). A loading solution including 2 M glycerol and 0.4 M sucrose is the most commonly employed in cryopreservation protocols (Sakai & Engelmann, 2007). The results obtained by Martinez-Montero et al., (2008) showed that modifying the composition of the loading solution improved viability according to electrolyte efflux test for cryopreserved sugarcane embryos.

This indicates the importance of carefully studying each step of the cryopreservation protocol to optimize survival. We hypothesized that increasing the number of OH groups present in the loading medium progressively decreased viability of control sugarcane somatic embryos, whereas there was an optimum in their number to achieve highest viability after cryopreservation (Fig. 6). It has been suggested that OH groups of sugars/polyalcohols replace water and interact with phospholipids forming hydrogen bonding with membrane phospholipids (Turner et al., 2001). This helps stabilizing cellular membranes during dehydration and cooling and helps maintaining membrane integrity and function through minimizing bilayer disruption and damages (Benson, 2007).

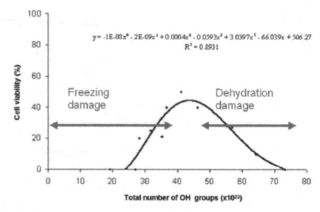

Fig. 6. Effect of the total number of OH groups of glycerol and sucrose on the cell viability of the cryopreserved sugarcane somatic embryos by droplet-vitrification procedure.

From a thermodynamic, kinetic, and structural point of view, the physico-chemical mechanism by which glycerol plus sucrose as co-solvent system can modulate the functionality of a given protein is very important (Baier & McClements, 2005; Ruan et al., 2003). However, the stabilization mechanism of these agents has been attributed to a protein preferential hydration mechanism, as proposed by Timasheff (1993) or to an osmotic stress (Parsegian et al., 1995) where, mathematically, the two mechanisms cannot be distinguished (Parsegian et al., 2000).

In a very well documented paper, Parsegian et al., (2000) indicated that there has been much confusion about the relative merits of different approaches, osmotic stress, preferential interaction (i.e. preferential hydration), and crowding, to describe the indirect effect of solutes on macromolecular conformations and reactions. The two first mechanisms (and crowding) cannot be distinguished as they are derived from the same solution theory. In the preferential hydration model proposed by Timasheff (1993), both the chemical nature and the size of the solute determine water exclusion from the protein surfaces. The osmotic stress emphasizes the role of the water that is necessarily included if solutes are excluded

Parsegian et al., 2000), dealing also with the movement of water molecules (Parsegian et al., 1995; Stanley et al., 2008).

Duan et al., (2001) observed that the hydroxyl groups present in the glucose units could contribute to protein-sugar interactions in aqueous solutions. These newly formed dipole-dipole interactions could form a hydrophilic layer around the protein units and therefore increase the dispersability of the protein through protein hydration and/or alter the intramolecular interactions in such a way that folding and even dissociation may be favoured (Semenova et al., 2002).

4. Stability assessment

Before using cryopreservation as an additional tool in the overall conservation strategy for any plant material, it is essential to verify that the cryopreservation protocol developed does not have any destabilizing effect and that the plants produced from cryopreserved explants are true to type (Harding, 2004). This author firstly provided a definition for "Cryobionomics" - a novel term describing the remodeled concept of genetic stability and the re-introduction of cryopreserved plants into the environment. Later, Cryobionomics is proposed as an approach to explore links between cryoinjury and genetic instability during and after cryopreservation (Harding et al., 2005, 2008b). There are an increasing number of reports indicating that no changes are observed in the material regenerated from cryopreservation (Engelmann, 1997). However, most of these experiments have been performed very soon after cryopreservation on a small number of individuals, often using material still cultured *in vitro* or after a very short period of growth in vivo and they concern mainly in vitro growth characteristics, or a limited number of biochemical or molecular markers. Only in a limited number of cases (e.g. Côte et al., 2000; Engelmann, 1997; Schäfer-Menuhr et al., 1997) have plants been grown in the field for a long period allowing the assessment of agronomic characteristics.

In the case of sugarcane, numerous experiments have been conducted to study the field behaviour of micropropagated plants (Feldmann et al., 1994; Flynn & Anderlini, (1990); Jackson et al., 1990; Lorenzo et al., 2001; Pena & Stay, 1997), uncovering the occurrence of rejuvenation phenomenons and of epigenetic changes. By contrast, only limited information is available concerning the stability of plants regenerated from cryopreserved material. RFLP analysis did not reveal any difference that could be attributed to cryopreservation between plants of one sugarcane variety produced from control and cryopreserved calluses (Eksomtramage et al., 1992) or cell suspensions (Chowdhury & Vasil 1993). Plants produced from control and cryopreserved shoot tips of one variety were similar as regards pattern of two isoenzymatic systems (Paulet et al., 1993) and six agronomic traits observed during their early growth in vivo (Gonzalez-Arnao et al, 1999).

Moreover, we published data on the field performance of sugarcane plants originating from cryopreserved material (Martinez-Montero et al., 2002b). The field performance of plants produced from embryogenic calluses of one sugarcane commercial hybrid cv. CP52-43 (CP43-64 x CP38-34, Canal Point, USA) cryopreserved using the protocol developed by Martinez-Montero et al. (1998) was evaluated over a period of 27 months by observing several agronomic parameters (Fig. 7). Similar observations were carried out simultaneously for comparison on plants produced from the same callus cultures, but which were not cryopreserved and on plants of the same variety originating from classical macropropagation.

Fig. 7. Regenerated plants from cryopreserved embryogenic sugarcane callus a) during acclimatization (42 days); b, c) after 12 months in the field of the stool; d) after 15 months in the field of the first ratoon.

Treatments were distributed following a randomised block experimental design including four repetitions per treatment. Experimental plots were 7.5 m long with 5 rows each. Intra-row spacing was 0.5 m. The commercial hybrid C266-70 (Co281 x POJ2878, INICA, Cuba) was planted as experiment border. Fertilization at the time of planting included 75 kg/ha urea; 50 kg/ha P_2O_5 and 50 kg/ha K_2O. Additionally, 75 kg/ha urea were supplied 3 months after planting. The following measurements were performed on 100 plants originating from cryopreserved calluses, control calluses and macropropagated plants:

- After 6 months of stool field growth: number of stems per clump (i.e. number of suckers produced from the original plant); stem diameter (cm); stem length (m).
- After 12 months of stool field growth: number of stems per clump; stem diameter (cm); stem length (m); fibre percentage (w/w); juice brix (i.e. mass of sugar (g dry matter) per 100 g of juice, expressed in Brix degree) ; pol (total sugar content) percentage in juice; pol percentage in cane; juice apparent purity tons of pol/ha.
- After cutting of stools and 15 months of field growth of the first ratoon: number of stems per clump; stem diameter (cm); stem length (m); single stem fresh mass (kg); fibre percentage (w/w); juice brix; pol percentage in juice; pol percentage in cane; juice apparent purity; tons of pol/ha.

The results of the evaluation of field grown sugarcane plants after different periods showed significant differences between treatments only during the first six months of field growth of sugarcane stools (Table 4). Stems produced from *in vitro* cultured materials, irrespective of their cryopreservation status, had a smaller diameter and a shorter height than those produced from macropropagated buds. These differences disappeared during the course of the experiment as they were not observed anymore after 12 months of stool field growth.

Parameter measured	Origin of stools			Typical Error
	Cryopreserved calluses	Control calluses	Buds isolated	
Stem diameter (cm)	1,51 b	1,45 b	1,82 a	0,12
Stem length (m)	0,57 b	0,49 b	0,85 a	0,10
Number of stems per meter	5,20 a	5,45 a	4,22 b	0,35

Table 4. Evaluation of several agronomic parameters after six months of field growth of sugarcane stools originating from cryopreserved calluses, control (non-cryopreserved) calluses and buds isolated from macropropagated plants. *Data in rows followed by the same letter are not statistically different (ANOVA, Tukey test, p < 0,05).*

No significant difference between treatments was observed for any of the parameters studied after 12 months of stool field growth and 15 months of field growth of the first ratoon (Tables 5 and 6). This study has demonstrated that the differences observed for several agronomic characters between stools originating from cryopreserved and control calluses, and macropropagated material after 6 months of field growth disappeared progressively with time, as no differences could be uncovered in stools after 12 months nor after 15 months of field growth of the first ratoon.

Only very few published reports on this topic deal with comparable number and duration components. No differences have been noted in the vegetative and floral development of several hundreds of palms regenerated from control and cryopreserved oil palm embryogenic cultures, but no detailed account of the observations made has been published (Engelmann, 1997). The most comprehensive study is the comparison of the field behaviour of banana plants regenerated from control and cryopreserved cell suspensions (Côte et al., 2000), which showed that two out of the eleven descriptors analyzed differed between control and cryopreserved material during the first culture cycle but that, similarly to our observations, these differences disappeared during the second culture cycle.

Parameter measured	Origin of stools			Typical Error
	Cryopreserved calluses	Control calluses	Buds isolated	
Stem diameter (cm)	2,58	2,51	2,66	0,09
Stem length (m)	1,85	1,90	1,70	0,11
Stem fresh weight (kg)	1,37	1,33	1,49	0,09
Stem number per meter	9,80	10,55	9,32	1,50
Agriculture recovery (t/ha)	111,88	116,93	115,72	2,62
Juice brix (°Brix)	22,07	21,22	24,69	1,87
Pol percentage in juice (%, w/w)	19,17	18,37	19,35	1,93
Fibre percentage (%, w/w)	11,55	11,23	11,52	0,43
Pol percentage in cane (%, w/w)	16,95	16,30	17,12	1,90
Agro-industrial recovery (t/ha)	18,96	19,06	19,81	1,51

Table 5. Evaluation of several agronomic parameters after 12 months of field growth of sugarcane stools originating from cryopreserved calluses, control (non-cryopreserved) calluses and buds isolated from macropropagated plants. *No statistical differences were found (ANOVA).*

An interesting result in our study is that plants regenerated from control and cryopreserved calluses displayed the same differences in comparison with those originating from macropropagated material. These differences are therefore not induced by cryopreservation but are due to the fact that both groups of plants originate from in vitro cultured material. It is indeed a well known phenomenon that tissue culture induces temporary changes in the behaviour of *in vitro* cultured plants during their early in vivo growth phase (Swartz 1991). Such changes can be induced by *in vitro* culture conditions including notably low light intensity, high humidity, limited gas exchanges, presence of high sucrose concentrations and growth regulators in the medium.

In the case of sugarcane, changes in the field behaviour of plants have been frequently observed after *in vitro* culture. Several authors have reported an increase in the number of new stems per clump (Flynn & Anderlini, 1990; Jackson, 1990; Perez et al., 1999), which generally induces a reduction in the stem diameter and mass. Peña & Stay (1997) stated that, with sugarcane *in vitro* culture stimulated growth and vigour, induced rejuvenation and generally improved agricultural performance. Many authors (Jimenez et al., 1991; Lorenzo et al., 2001; Lourens. & Martin 1987; Taylor et al., 1992) indicate that such differences disappear during the course of field growth and the first clonal multiplication, as observed in our experiments.

Parameter measured	Origin of stools			Typical Error
	Cryopreserved calluses	Control calluses	Buds isolated	
Stem diameter (cm)	2,62	2,59	2,65	0,10
Stem length (m)	1,93	1,98	2,04	0,16
Stem fresh weight (kg)	1,50	1,51	1,54	0,21
Stem number per meter	9,77	9,65	10,09	2,20
Agriculture recovery (t/ha)	122,10	121,42	129,48	5,12
Juice brix (°Brix)	23,22	23,02	22,98	1,79
Pol percentage in juice (%, w/w)	20,45	20,07	19,95	1,85
Fibre percentage (%, w/w)	12,9	12,17	12,5	0,35
Pol percentage in cane (%, w/w)	17,81	17,62	17,45	1,80
Agro-industrial recovery (t/ha)	21,75	21,35	22,59	1,51

Table 6. Evaluation of several agronomic parameters after 15 months of field growth of the first sugarcane ratoon originating from cryopreserved calluses, control (non-cryopreserved) calluses and buds isolated from macropropagated plants. *No statistical differences were found (ANOVA).*

In conclusion, the results obtained in our study validate the cryopreservation protocol developed by Martinez-Montero et al. (1998) for embryogenic calluses. Cryopreservation of embryogenic calluses will thus be incorporated in the scheme established by the Centro de

Bioplantas for mass production of *in vitro* sugarcane plants by means of somatic embryogenesis.

5. Cryopreservation protocols for pineapple

Pineapple is vegetatively propagated and crosses between varieties produce botanical seeds. However, these seeds are highly heterozygous and therefore of limited interest for the conservation of specific gene combinations. Cryopreservation of apices is the most relevant strategy for long-term conservation of vegetatively propagated crops, since true to type; virus free plants can be regenerated directly from cryopreserved apices (Lynch et al., 2007).

Vitrification and encapsulation-dehydration techniques have been widely applied for successfully cryopreserve apices of a large number of different crops which do not require sophisticated equipment for freezing and produce high recovery rates with a wide range of materials (Engelmann, 2010, 2011). Most vitrification protocols use a loading treatment and a stringently timed dehydration with PVS (Benson, 2008b). This two-step procedure has allowed tissues to be more tolerant to osmotic stress and to resist the chemical toxicity induced by the highly concentrated cryoprotective solutions. Exposure duration to PVS is usually not longer than 2 h (Thinh et al., 1999).

The encapsulation-dehydration method for cryopreservation is based on the fact that encapsulation protects the explants and preculture in medium enriched with osmoticum makes them tolerant to air drying (Fabre & Dereuddre, 1990). Preculture in 0.75 M sucrose and desiccation to about 25% water content in beads (fresh weight basis) are the most common conditions used (Gonzalez-Arnao et al., 1996). The application of vitrification solutions to dehydrate encapsulated cells or shoot tips has also been successfully applied to several species (Engelmann, 2004).

5.1 Apices

The first successful result related with the cryopreservation of pineapple (*Ananas comosus*) apices was reported by our group in Cuba (Gonzalez-Arnao et al., 1998b). The encapsulation and vitrification techniques were experimented for freezing apices of pineapple *in vitro* plantlets. Positive results were achieved using vitrification only (Fig. 8). Optimal conditions included a 2 day preculture of apices on medium supplemented with 0.3M sucrose, loading treatment for 25 min in medium with 0.75M sucrose + 1M glycerol and dehydration with PVS2 vitrification solution at 0°C for 7 h before rapid immersion in liquid nitrogen. This method resulted in ranged survival (25-65%) depending of the genotypes. Recovery of cryopreserved apices took place directly, without transitory callus formation.

The negative results after cryopreservation of pineapple apices by encapsulation-dehydration technique can be related to the high sensitivity of pineapple apices to sucrose and dehydration. Pregrowth in media with sucrose concentrations higher than 0.5M was detrimental to survival and a prolonged treatment in 0.5M sucrose was required to improve survival after desiccation, but it was not sufficiently to obtain survival of apices after freezing. The viability loss observed after freezing may be due to the crystallization of remaining freezable water upon freezing. This detrimental crystallization might be avoided by slowly freezing of the encapsulated apices, which would result in removing the remaining freezable water by means of freeze-induced dehydration. Several plant materials cryopreserved by encapsulation-dehydration technique have required slow freezing regime to achieved optimal survival (Engelmann, 2010).

In another set of experiments, our group obtained positive results again only with vitrification (Martinez-Montero et al., 2005). The best protocol comprised a 2-d preculture on semi-solid MS medium supplemented with 0.3 M sucrose, a loading treatment in liquid medium containing 0.4 M sucrose + 2 M glycerol for 25 min, and dehydration for 7h at 0°C with PVS3 before immersion into liquid nitrogen. The highest survivals of apices were: Smooth Cayenne (45%), Cabezona (33%) and Red Spanish (25%).

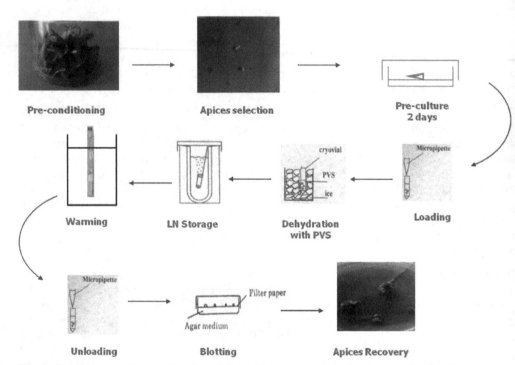

Fig. 8. Cryopreservation protocol established for pineapple apices using vitrification technique.

However, contrary to most vitrification reports, pineapple apices required a prolonged exposure (7 h) to the vitrification solutions (Engelmann, 2010). This result is probably due to the large size, and compact structure of the pineapple apices employed in our experiments with success (Fig. 9): the apices were around 3mm long, and comprised the apical dome tightly covered by 2-3 leaf primordial with a very thick cuticle. Long treatment durations were needed for the vitrification solution to sufficiently dehydrate these very compact structures.

Cryopreservation of Tropical Plant Germplasm with Vegetative Propagation – Review of Sugarcane
Saccharum spp.) and Pineapple (Ananas comusus (L.) Merrill) Cases
147

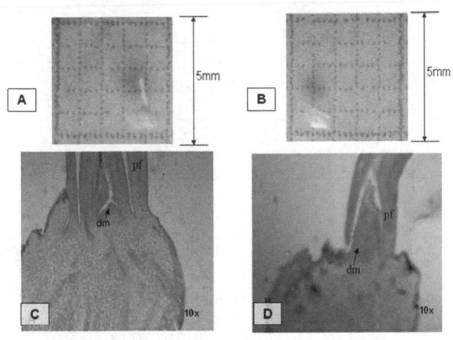

Fig. 9. Dissected pineapple (*Ananas comosus* L. Merrill cv. MD2) shoot tips as viewed by stereo-microscope (A, B); and by light microscopy 10x (C, D). Dissected apices type I (A, C) with apical dome (dm) and 3-4 primordial leaves (pf) and mechanical damaged apices type II (B, D) with one primordial leaf used as controls.

5.1.1 Extension of vitrification protocol

In the case of pineapple, it is believe as symbol in the province of Ciego de Avila (Cuba) due to great cultivated areas dedicated to this crop. In our Institution (Bioplantas Centre) is located the unique field collection of pineapple germplasm in the country. However, this field genebank is prone to disease, or damage through natural disaster and need very high maintenance. For this reason, the cryopreservation of apices obtained from vitroplants could constitutes the most relevant strategy for long-term conservation of pineapple germplasm, since true to type and virus-free plants can be regenerated directly from cryopreserved apices (Martinez-Montero et al., 2005).

The successful application of the vitrification protocol for nine accessions of the *in vitro* collection at Bioplantas Centre was accomplished with the following conditions: type of shoot tip (consisted in meristematic dome area and 3-4 primordial leaves with 2,5 – 3 mm in size); 0,3 mol.L⁻¹ sucrose preculture during 2 days; application of the loading solution (0,4 mol.L⁻¹ sucrose + 2 mol.L⁻¹ glycerol) during 25 min at 25°C; dehydration with plant vitrification solution number three (PVS3: 50% w/v glycerol + 50% w/v sucrose) during 7 hours at 0°C. The results per accessions expressed as percentage of recovery before (-LN) and after (+LN) cryopreservation for six apices per replicate, four replicates per treatment and each experiment was repeated three times are showed in Table 7.

	Survival (%)	
	-LN	+LN
Cayenne of Puerto Rico	80.2	65.5
Perolera	49.9	33.8
Smooth Cayenne of Serrana	50.3	25.3
Cabezona	61.5	27.9
Piña Blanca of Caney	57.9	24.7
P3R5	53.1	20.0
Red Spanish of Caney	45.5	12.1
MD2	80.1	60.2
Bromelia sp.	33.1	6.3

Table 7. Effect of vitrification protocol on survival of apices from eight pineapple accessions and one related specie (*Bromelia* sp.) before (-LN) and after cryopreservation (+LN).

5.1.2 Optimization of methodology for pineapple apices

Further modifications to the procedures might be useful in order to reduce the exposure duration to PVS and achieve higher survival after cooling. Therefore, the objective was to develop a more effective cryopreservation protocol using both vitrification and encapsulation/vitrification. As previously we reported (Gonzalez-Arnao et al., 1998b; Martinez-Montero et al., 2005), pineapple apices are sensitive to sucrose and dehydration exposures. A progressive treatment increasing sucrose concentrations was effective at enhancing their tolerance to dehydration and cooling.

Proline has been shown to have a beneficial effect in several cryopreservation protocols (Luo & Reed, 1997; Rasmunsen et al., 1997; Rudolf & Crowe, 1985; Thierry et al., 1999). In our experiments we confirmed that a 2-d progressive preculture in a mixture of sucrose and proline improved the results obtained after cooling in comparison with sucrose alone. This modification in pregrowth also considerably reduced the required dehydration time in PVS. As previously reported, apices of Puerto Rico variety treated for 2 d in 0.3 M sucrose needed an extended exposure (7 h) to PVS2 at 0°C to achieve high levels of survival (65 %) after cryopreservation. However, following the same vitrification approach, higher survival (72%) was obtained using the best pretreatment in sucrose-proline and only 30 min of exposure to PVS2. As regards PVS3, we also demonstrated that 30 min or 1 h of dehydration were also enough to obtain higher survival (76 %) after cryopreservation.

The role of proline has been associated with its ability to act as source of nitrogen and carbon for reparative post-stress processes (Rabbe & Lova, 1984), to increase the non-freezable fraction of water (Rasmussen et al, 1997), to inhibit membrane mixing and to stabilize proteins during dehydration and freezing (Rudolf & Crowe, 1985). On the other hand, the combination of chemical cryoprotectants may also improve the response of tissues to cryopreservation in comparison with the application of one chemical alone. As reported for wasabi apices, a mixture of glycerol with sucrose was more effective at enhancing their tolerance to dehydration and deep cooling than a preculture in sucrose alone (Sakai et al, 2000).

Dehydration at 0°C instead of 25°C for both vitrification solutions gave better results, as previously reported for pineapple apices by our group. This low temperature reduces the toxicity of the vitrification solutions and increases the potential period of exposure (Withers & Engelmann, 1997). In all our cryopreservation experiments, dehydration with PVS3 at 0°C gave higher recovery rates compared with PVS2, even for diverse genotypes. The encapsulation-vitrification method using PVS3 gave greater survival than the vitrification procedure. Additionally, the manipulation of encapsulated apices permits handling large quantities of material that from the practical point of view is more convenient. These results corroborated that encapsulation-vitrification may also be very useful for cryopreserving desiccation-sensitive germplasm such as pineapple, that could not be successfully cryopreserved using an encapsulation-dehydration approach (Gonzalez-Arnao et al., 2003).

The cryopreservation protocol presented here improved survival and shortened the process compared with previous protocols from our group (table 8). The optimal conditions involved the encapsulation of pineapple apices in calcium alginate (3%) beads, followed by a 2-d progressive preculture in liquid medium with 0.16 M sucrose + 0.3 M proline for 24 h, then 0.3 M sucrose + 0.3 M proline for 24h, a loading treatment for 25 min in 0.75 M sucrose + 1M glycerol solution at room temperature and dehydration for 60 min with PVS3 at 0°C before rapid immersion into liquid nitrogen.

| Variety | Vitrification Solution | Regrowth (%) | | | |
| | | Encapsulation-Vitrification | | Vitrification | |
		-LN	+LN	-LN	+LN
MD-2					
	PVS2	75±2.7 [a]	39±1.9 [d]	68±1.0 [b]	46±2.0 [c]
	PVS3	76±1.9 [a]	54±2.8 [c]	72±2.0 [b]	48±1.1 [d]
Puerto Rico	PVS2	91±1.3 [a]	83±1.8 [c]	88±1.5 [b]	72±2.0 [d]
	PVS3	93±2.2 [a]	83±1.1 [b]	88±1.4 [a]	76±0.7 [c]

Different letters within rows imply significant differences according to Tukey test ($P<0.05$) and each PVS.

Table 8. Comparison of the encapsulation-vitrification and vitrification procedures on regrowth (%) of pineapple apices after dehydration with PVS2 or PVS3 solutions at 0°C.

5.2 Embryogenic callus

The results of some studies indicated *Fusarium subglutinans* isolates cause fusariose which constitutes the most serious pineapple disease and causes losses as high as 80% of marketable pineapple fruit. It produces phytotoxins in culture that were phytotoxic on calluses (Jin et al., 1996; Kaur et al., 1987) and the correlation between pineapple variety and the toxicity of culture filtrates suggests that filtrates could be used to screen in vitro for disease resistance (Borras et al., 2001). Therefore, the cryopreservation of pineapple calluses can provide a means of effective source when in vitro screening of germplasm for fusariose disease would be attempted. Storing calluses in liquid nitrogen could preserve their regeneration capacity and limits the risk of somaclonal variation, which increases with culture duration.

A simplified freezing protocol mentioned before for sugarcane embryogenic calluses (Martinez-Montero et al., 1998) was used for pineapple calluses of the genotypes "Smooth Cayenne" and "Perolera" (Martinez-Montero et al., 2005). For cryopreservation experiments, 15 to 20 day-old calluses, about 3 to 6 mm in diameter, were employed. Calluses were pretreated with a cryoprotective solution containing 5 to 20 % (v/v) dimethylsulfoxide (DMSO) and 0.5M sucrose for 1h at 0°C. After freezing calluses were transferred directly to recovery medium (MS medium supplemented with dicamba:BAP (2.5:0.5 mg.L^{-1}) and citric acid (0.1 mg.L^{-1})). The survival, evaluated 45 days after thawing, corresponded to the percentage of calluses which had increased in size during the recovery period.

As results, the survival of calluses after pretreatment was high and similar for the two genotypes studied (Table 8). It decreased only with 20% of DMSO. After freezing in liquid nitrogen, survival was achieved with 10 and 15% DMSO only and was highest with 15% DMSO (57-67%). Re-growth of successfully cryopreserved calluses was very rapid and they increased in size during the recovery period. With this work, the application of a simplified freezing protocol achieved survival from used pineapple genotypes and this confirm that our previously simplified freezing protocol for sugarcane (Martinez-Montero et al., 1998) can be used wider to others species.

DMSO (%)	Sucrose (M)	Survival (%)			
		Smooth Cayenne		Red Spanish	
		-LN	+LN	-LN	+LN
5		85±11 a	0 c	87±6 a	0 c
10	0.5	81±8 a	25±3 b	85±6 a	30±4 b
15		75±12 a	57±7 a	80±6 a	65±6 a
20		55±6 b	0 c	60±4 b	0 c

Table 9. Effect of cryoprotective solution (DMSO + sucrose) on the survival rate (%) before (-LN) and after (+LN) application of cryopreservation protocol for calluses of two pineapple genotypes. *Values represent means of 50 samples from three replicate experiments, ± SE. Means within columns followed by the same letter are not significantly different (ANOVA p<0,05 Tukey,). Data were transformed for statistical analysis in accordance with x'= 2 arcsine ((x/100)^{0,5}) for percentage of survival.*

6. Conclusions

For vegetatively propagated species, cryopreservation has a wide applicability both in terms of species coverage, since protocols have been successfully established for root and tubers, fruit trees, ornamentals and plantation crops, both from temperate and tropical origin and in terms of numbers of genotypes/varieties within a given species. With a few exceptions, vitrification-based protocols have been employed. It is also interesting to note that in many cases, different protocols can be employed for a given species and produce comparable results. Survival is generally high to very high. Regeneration is rapid and direct, and callusing is observed only in cases where the technique is not optimized. Different reasons can be mentioned to explain these positive results. The meristematic zone of apices, from which organised growth originates, is composed of a relatively homogenous population of small, actively dividing cells, with little vacuoles and a high nucleocytoplasmic ratio. These characteristics make them more susceptible to withstand desiccation than highly vacuolated and differentiated cells. As mentioned earlier, no ice formation takes place in vitrification-based procedures, thus avoiding the extensive damage caused by ice crystals which are formed during classical procedures. The whole meristem is generally preserved when vitrification-based techniques are employed, thus allowing direct, organised regrowth. By contrast, classical procedures often lead to destruction of large zones of the meristems, and callusing only or transitory callusing is often observed before organised regrowth starts. Other reasons for the good results obtained are linked with tissue culture protocols. Many vegetatively propagated species successfully cryopreserved until now are cultivated crops, often of great commercial importance, for which cultural practices, including *in vitro* micropropagation, are well established. In addition, *in vitro* material is "synchronized" by the tissue culture, and pregrowth procedures and relatively homogenous samples in terms of size, cellular composition, physiological state and growth response are employed for freezing, thus increasing the chances of positive and uniform response to treatments. Finally, vitrification-based procedures allow using samples of relatively large size (shoot tips of 0.5 to 2–3 mm), which can regrow directly without any difficulty. Cryopreservation techniques are now operational for large-scale experimentation in an increasing number of cases. In view of the wide range of efficient and operationally simple techniques available, any vegetatively propagated species should be amenable to cryopreservation, provided that the tissue culture protocol is sufficiently operational for this species.

7. Acknowledgment

The authors thank the Food and Agriculture Organization of the United Nations (FAO), the International Plant Genetic Resources Institute (IPGRI, actually Bioversity International) and the International Foundation for Science (IFS) for partly funding the research programme.

8. References

Anchordoguy, T.J.; Cecchini, C.A.; Crowe, J.H.; Crowe, L.M. (1991) Insights into the cryoprotective mechanism of dimethyl sulphoxide for phospholipid bilayers. *Cryobiology* 28(5), (October 1991), 467-473. ISSN: 0011-2240

Arakawa, T.; Carpenter, J.F.; Kita, Y.A.; Crowe, J.H. (1990) The basis for toxicity of certain cryoprotectants: An hypothesis. *Cryobiology* 27(4), (August 1990), 401-15. ISSN: 0011-2240

Aronen, T.S.; Krajnakova, J.; Haggman, H.M., Ryynannen, L.A. (1999) Genetic stability of cryopreserved embryogenic clones of white spruce (*Picea glauca*). *Plant Cell Rep.* 18, 948-953. ISSN: 0721-7714

Ashmore, S.E. (1997) *Status reports on the development and application of in vitro techniques for the conservation and use of plant genetic resources.* International Plant Genetic Resources Institute, ISBN 92-9043-339-6, Rome, Italy.

Ausborn, M.; Schreier, H.; Brezesinski, G.; Fabian, H.; Meyer, H.W.; Nuhn, P. (1994) The protective effect of free and membrane bound cryoprotectants during freezing and freeze-drying of liposomes. *J Control Release* 30, 105-116. ISSN: 0168-3659

Baier, S.K. & McClements, D.J. (2005) Influence of cosolvent systems on the gelation mechanism of globular protein: Thermodynamic, kinetic, and structural aspects of globular protein gelation. *Comprehensive Reviews in Food Science and Food Safety* 4, 43-54. ISSN: 1541-4337

Barraco, G.; Sylvestre, I.; Engelmann, F. (2011) Comparing encapsulation-dehydration and droplet-vitrification for cryopreservation of sugarcane (*Saccharum* spp.) shoot tips. *Scientia Horticulturae*, 130(1), (August 2011), 320-324. ISSN: 0304-4238

Benson, E.E. (2008a) Cryopreservation of phytodiversity: A critical appraisal of theory and practice. *Critical Reviews in Plant Sciences* 27(3), 141-219. ISSN: 0735-2689

Benson, E.E. (2008b) Cryopreservation theory. In: *Plant Cryopreservation: A Practical Guide.* Reed, B.M. Ed., pp. 15–32. Chapt. 2. Springer, ISBN 978-0-387-72275-7, New York, USA.

Benson, E.E., Lynch, P.T., Jones, J. (1992) Variation in free radical damage in rice cell suspensions with different embryogenic potentials. *Planta* 188: 296–305. ISSN: 0032-0935

Borras, O.; Santos, R.; Matos, A.; Cabral, R.; Arzola, M. (2001) A first attempt to use a *Fusarium subglutinans* culture filtrate for the selection of pineapple cultivars resistant to fusariose disease. *Plant Breed.* 120: 435-438. ISSN: 0179-9541

Chen, T.H.H.; Kartha, K.K.; Constable, F.; Gusta, L.V. (1984) Freezing characteristics of cultured *Catharanthus roseus* (L.) G. Don cells trated with dimethylsulfoxide and sorbitol in relation to cryopreservation. *Plant Physiol.* 75, 720-725. ISSN: 0032-0889

Chowdhury, M.K.U. & Vasil, I.K. (1993) Molecular analysis of plant regenerated from embryogenic cultures of hybrid sugarcane cultivars (*Saccharum* spp.). *Theor. Appl. Genet* 86, 181-188. ISSN: 0040-5752

Côte, F.X.; Goué, O.; Domergue, R.; Panis, B.; Jenny, C. (2000) In-field behavior of banana plants (*Musa* spp.) obtained after regeneration of cryopreserved embryogenic cell suspensions. *Cryo-Letters* 21, 19-24. ISSN: 0143-2044

Cyr, D.R. (2000) Cryopreservation: roles in clonal propagation and germplasm conservation of conifers. In: *Cryopreservation of tropical plant germplasm - Current research progress and applications.* Engelmann, F.; Takagi, H., Eds. (261-268) Tsukuba: JIRCAS; Rome: IPGRI. ISBN 92-9043-428-7, Rome, Italy.

Duan, X.Q.; Hall, J.A.; Nikaido, H.; Quicho, F.A. (2001) Crystal structures of the maltodextrin/maltose-binding protein complexed with reduced oligosaccharides: flexibility of tertiary structure and ligand binding. *Journal of Molecular Biology* 306:5, (9 March 2001), 1115-1126. ISSN: 0022-2836

Dumet, D.; Block,W.; Worland, R.; Reed, B.M.; Benson, E.E. 2000. Profiling cryopreservation protocols for *Ribes ciliatum* using differential scanning calorimetry. *Cryo-Letters* 21: 367–378. ISSN: 0143-2044

Eksomtramage T.; Paulet F.; Guiderdoni E.; Glaszmann J.C.; Engelmann F. (1992) Development of a cryopreservation process for embryogenic calluses of a commercial hybrid of sugarcane *(Saccharum* sp.) and application to different varieties. *Cryo-Letters* 13, 239-52. ISSN: 0143-2044.

Engelmann, F. & Takagi, H. Eds. (2000) *Cryopreservation of tropical plant germplasm - Current research progress and applications.* Tsukuba: JIRCAS; Rome: IPGRI; ISBN 92-9043-428-7, Rome, Italy.

Engelmann, F. (1997) Importance of desiccation for the cryopreservation of recalcitrant seed and vegetatively propagated species. *Plant Genet. Res. Newsletter* 112, 9-18. ISSN: 1020-3362

Engelmann, F. (1997) *In vitro* conservation methods. In: *Biotechnology and Plant Genetic Resources Conservation and Use.* BV Ford-Lloyd, JH Newburry & JA Callow (Eds). pp. 119-161. ISBN: 0851991424, CABI, Wallingford.

Engelmann, F. (2000) Importance of cryopreservation for the conservation of plant genetic resources. In: *Cryopreservation of tropical plant germplasm - Current research progress and applications.* Engelmann, F.; Takagi, H., Eds. (8-20) Tsukuba: JIRCAS; Rome: IPGRI; ISBN 92-9043-428-7, Rome, Italy.

Engelmann, F. (2004) Plant cryopreservation: progress and prospects. *In Vitro Cell Dev Biol Plant* 40, 427–433. ISSN 1054-5476

Engelmann, F. (2010) Use of biotechnologies for the conservation of plant biodiversity. *In Vitro Cell Dev Biol Plant* DOI 10.1007/s11627-010-9327-2, ISSN 1054-5476

Engelmann, F. (2011) Cryopreservation of Embryos: An Overview. In: *Plant Embryo Culture: Methods and Protocols, Methods in Molecular Biology.* Trevor, A.T. & Yeung, E.C. Eds. Vol. 710, DOI 10.1007/978-1-61737-988-8_13, ISBN 978-1-61737-987-1, Springer Science+Business Media, LLC 2011.

Esterbauer, H.; Zollner, H.; Schauer, R.J. (1988) Hydroxyalkenals: Citotoxics products of lipid peroxidtion. In: *ISI Atlas of Sciences: Biochemistry* 311-317. ISBN. 0-941708-00-4

Fabre, J. & Dereuddre, J. (1990) Encapsulation-dehydration: A new approach to cryopreservation of *Solanum* shoot tips. *Cryo-Letters* 11, 413 – 426. ISSN: 0143-2044

Fahy, G.M.; Lilley, T.H.; Linsdell, H.; Douglas, M.S.J.; Meryman, H.T. (1990) Cryoprotectant toxicity and cryoprotectant toxicity reduction: in search of molecular mechanisms. *Cryobiology* 27(3), (June 1990), 247-68. ISSN: 0011-2240

Fahy, G.M.; McFarlane, D.R.; Angell, C.A.; Meryman, H.T. (1984) Vitrification as an approach to cryopreservation. *Cryobiology,* 21(4), (August 1984), 407-426. ISSN: 0011-2240

Feher, A.; Pasternak, T.P.; Dudits, D. (2003) Transition of somatic cells to an embryogenic state. *Plant Cell, Tissue and Organ Culture* 74, 201-228. ISSN: 0167-6857

Feldmann, P.; Sapotille, J.; Gredoire, P.; Rott, P. (1994) Micropropagation of sugarcane. In: *In vitro culture of tropical plants.* C Teisson (ed). CIRAD, ISBN 2-87614-162-0, Montpellier, pp. 15-17.

Finkle, B.J. & Ulrich, J.M. (1979) Effect of cryoprotectants in combination on the survival of frozen sugarcane cells. *Plant Physiology* 63, 598-604. ISSN: 0032-0889

Finkle, B.J. & Ulrich, J.M. (1982) Cryoprotectant removal as a factor in the survival of frozen rice and sugarcane cells. *Cryobiology* 19(3), (June 1982) 329-335. ISSN: 0011-2240

Finkle, B.J. ; Zavala, M.E.; Ulrich, J.M. (1985) Cryoprotective compounds in the viable freezing of plant tissues. In: *Cryopreservation of plant cells and organs.* Kartha, K.K. Ed., 243-267, ISBN 0-8493-6102-8, CRC Press, Boca Raton. USA.

Fleck, R.A.; Day, J.G.; Clarke, K.J.; Benson, E.E. (1999) Elucidation of the metabolic and structural basis for the cryopreservation recalcitrance of *Vaucheria sessilis, Xanthophyceae. Cryo-Letters* 20, 271–282. ISSN: 0143-2044

Flynn, L. & Anderlini, T. (1990) Disease incedence and yield performance of tissue culture-generated seed cane over the crop cycle in Lousiana. *Journal of American Society of Sugar Cane Technologists* 10, 113. ISSN 1991-8178

Gnanapragasam, S. & Vasil, I.K. (1990) Plant regeneration from a cryopreserved embryogenic cell suspension of a commercial sugarcane hybrid (*Saccharum* sp.) *Plant Cell Rep.* 9, 419-423. ISSN: 0721-7714

Gnanapragasam, S. & Vasil, I.K. (1992) Cryopreservation of immature embryos, embryogenic callus and cell suspension cultures of gramineous species. *Plant Science* 83, 205-215. ISSN: 0168-9452

Gonzalez-Arnao, M.T.; Engelmann F.; Huet C.; Urra C (1993) Cryopreservation of encapsulated apices of sugarcane: Effect of freezing procedure and histology. *Cryo-Letters* 14, 303-308. ISSN: 0143-2044

Gonzalez-Arnao, M.T.; Engelmann. F.; Urra-Villavicencio. C.; Morenza, M.; Rios, A. (1998a) Cryopreservation of citrus apices using the encapsulation-dehydration technique. *Cryo-Letters* 19,177-182. ISSN: 0143-2044

Gonzalez-Arnao, M.T.; Juarez, J.; Ortega, C.; Navarro, L.; Duran-Vila, N. (2003) Cryopreservation of ovules and somatic embryos of citrus using the encapsulation-dehydration technique. *Cryo-Letters* 24 (2), 85-94. ISSN: 0143-2044

Gonzalez-Arnao, M.T.; Moreira, T.; Urra, C. (1996) Importance of pregrowth with sucrose and vitrification for the cryopreservation of sugarcane apices using encapsulation-dehydration. *Cryo-Letters* 27, 141-148. ISSN: 0143-2044

Gonzalez-Arnao, M.T.; Ravelo, M.M.; Urra-Villavicencio, C.; Martinez-Montero, M.M.; Engelmann, F. (1998b) Cryopreservation of pineapple (*Ananas comosus*) apices. *Cryo-Letters* 19, 375-382. ISSN: 0143-2044

Gonzalez-Arnao, M.T.; Urra, C.; Engelmann, F.; Ortiz, R.; de la Fe, C. (1999) Cryopreservation of encapsulated sugarcane apices: Effect of storage temperature and storage duration. *Cryo-Letters* 20, 347-352. ISSN: 0143-2044

Grune, T.; Michel, P.; Sitte, N.; Eggert, W.; Albrecht-Nebe, H.; Esterbauer, H.; Siems, W.G. (1997) Increased levels of 4-hydroxynonenal modified proteins in plasma of children with autoimmune diseases. *Free Radical Biology and Medicine* 23 (3), (November 1996), 357-360. ISSN: 0891-5849

Guy, C. (1999) Molecular responses of plants to cold shock and cold acclimation. *J Mol Microbiol Biotechnol.* 1(2), 231-242. ISSN: 1464-1801

Harding, K. (1999) Stability assessments of conserved plant germplasm. In: *Plant Conservation & Biotechnology.* Chapt. 7. Benson, E.E. Ed., pp. 97–107, Taylor and Francis Ltd., ISBN 0-7484-0746-4, London, UK.

Harding, K. (2004) Genetic integrity of cryopreserved plant cells: A review. *Cryo-Letters* 25, 3-22. ISSN: 0143-2044

Harding, K.; Johnston, J.; Benson, E. E. (2005) Plant and algal cell cryopreservation: issues in genetic integrity, concepts in 'Cryobionomics' and current European applications. In: *Contributing to a Sustainable Future*. Benett, I.J.; Bunn, E.; Clarke, H.; McComb, J. A., Eds., (112–119) Proc. Australian Branch IAPTC & B, Perth, Western Australia.

Ishikawa, M.; Ide H.; Price, W.S.; Arata, Y.; Kitashima, T. (2000) Freezing behaviors in plant tissues as visualized by NMR microscopy and their regulatory mechanisms. In: *Cryopreservation of tropical plant germplasm - Current research progress and applications*. Engelmann, F.; Takagi, H., Eds. (22-35) Tsukuba: JIRCAS; Rome: IPGRI. ISBN 92-9043-428-7, Rome, Italy.

Jackson, W.; Wagnespack, H.; Richard, C.; Garrison, D.; Lester, W. (1990) CP65-357 Kleentek test in Lousiana. *Journal of the American Society of Sugar Cane Technologist* 21, 10-13. ISSN 1991-8178

Jian, L.C.; Sun, D.L.; Sun, L.H. (1987) Sugarcane callus cryopreservation. In: *Plant cold hardiness*. Li, P.H. Ed., 323-337, Alan R. Liss, Inc., ISBN 0-8451-1084-8, New York.

Jiménez, E.; Pérez, J.; Gil, V.; Herrera, J.; García, Y.; Alfonso, E. (1995) Sistemas para la propagación de caña de azúcar. In: *Avances de la Biotecnología Moderna* Estrada M, Riego E, Limonta E, Tellez P, Fuentes J (Eds).. Elfos Scientiae. Cuba 3:11.2

Jin, H.; Hartman, G.L.; Nickell, C.D.; Widholm, J.M. (1996) Phytotoxicity of culture filtrate from *Fusarium solani*, the causal agent of Soybean Sudden Death Syndrome. *Plant Dis*. 80: 922-927. ISSN: 0191-2917

Joyce, S.M.; Cassells A.C.; Jain, M.S. (2003) Stressand aberrant phenotypes *in vitro* culture. *Plant Cell Tissue and Organ Culture* 74, 103-21. ISSN:0167-6857

Kaczmarczyk, A.; Shvachko, N.; Lupysheva, Y.; Hajirezaei, M.R.; Keller, E.R.J. (2008) Influence of altering temperature preculture on cryopreservation results for potato shoot tips. *Plant Cell Rep*. 27: 1551-1558. ISSN: 0721-7714

Kartha, K.K. & Engelmann F. (1994) Cryopreservation and germplasm storage. In: *Plant cell and tissue culture*. Vasil, I.K. & Thorpe, T.A. Eds. (195–230) Kluwer, ISBN 978-0-7923-2493-5, Dordrecht.

Kartha, K.K.; Fowke, L.C.; Leung, N.L.; Caswell, K.L.; Hakman I. (1988) Induction of somatic embryos and plantlets from cryopreserved cell cultures of white spruce (*Picea glauca*). *J. Plant Physiol* 132, 529-39. ISSN: 0176-1617

Kaur, G.; Singh, U.S.; Garg, G.K. (1987) Mode of action of toxin isolated from *Fusarium oxysporum* f. sp. ciceri. *Indian Phytopathol* 40: 76-84. ISSN 0367-973X

Kinoshita, K.; Li, S.J.; Yamazaki, M. (2001) The mechanism of the stabilization of the hexagonal II (HII) phase in PE membranes in the presence of low concentrations of dimethyl sulfoxide. *Eur Biophysics J* 30, 207-220. ISSN 0175-7571

Klimazewska, K.; Ward, C.; Cheliak, W.M. (1992) Cryopreservation and plant regeneration from embryogenic cultures of larch (*Larix* x *eurolepis*) and black spruce (*Picea mariana*). *J Exp Bot* 43, 73-79. ISSN: 0022-0957

Krishnamurthy, K.V.; Kishnaraj, R.; Chozavendam, R.; Samuel, C.F. (2000) The programme of cell death in plants and animals. A comparison. *Curr Sci* 79, 1169-1181. ISSN: 0011-3891

Krishnaraj, S. & Vasil, I.K. (1995) Somatic embryogenesis in herbaceous monocots. In: *In vitro embryogenesis in plants. Current Plant Science and Biotechnology in Agriculture*,

Thorpe, T.A. Ed. Kluwer Academic Publishers; 20, 417-469. ISBN: 0-7923-3149-4, Dordrecht , Netherlands.

Kristensen, M.M.H.; Find, J.I.; Floto, F.; Moller, J.D.; Norgaard, J.V.N.; Krogstrup, P. (1994) The origin and development of somatic embryos following cryopreservation of an embryogenic suspension culture of *Picea sitchensis*. *Protoplasma* 182, 65-70. ISSN: 0033-183X

Leprince, O.; Hendry, G.A.F.; McKersie, B.D. (1993) The mechanisms of desiccation tolerance in developing seeds. *Seed Science Research*, 3: 231-246. ISSN: 0960-2585

Lorenzo, J.C.; Ojeda, E.; Espinosa, A.; Borroto, C. (2001) Field performance of temporary bioreactor-derived sugarcane plants. *In Vitro Cell Dev. Biol.-Plant*. 37, 803-806. ISSN 1054-5476

Lourens, A.G. & Martin, F.A. (1987) Evaluation of *in vitro* propagated sugarcane hybrids for somaclonal variation. *Crops Science* 27, 793-796. ISSN: 0011-183X

Lunn, J.E. & MacRae, E. (2003) New complexities in the synthesis of sucrose. *Current Opinion in Plant Biology* 6, 208-224. ISSN: 1369-5266

Luo, J. & Reed, B.M. (1997) Abscisic acid-responsive protein bovine serum albumin and proline pretreatments improve recovery of *in vitro* currant shoot tips and callus cryopreserved by vitrification. *Cryobiology* 34:240-250. ISSN: 0011-2240

Lynch, P.T.; Benson, E.E.; Harding, K. (2007) Climate change: the role of ex situ and cryo-conservation in the future security of economically important, vegetatively propagated plants. *Journal of Horticultural Science & Biotechnology* 82 (2), 157-160. ISSN: 1462-0316

Maddox, A.D.; Gonsalves, F.; Shields, R. (1983) Successful preservation of suspension cultures of three *Nicotiana* species at the temperature of liquid nitrogen. *Plant Science Letters* 28, 157-162. ISSN: 0304-4211

Martinez-Montero, M.E.; Gonzalez-Arnao, M.T.; Borroto-Nordelo, C.; Puentes-Diaz, C.; Engelmann, F. (1998) Cryopreservation of sugarcane embryogenic callus using a simplified freezing process. *Cryo-Letters*, 19, 171-176. ISSN: 0143-2044

Martinez-Montero, M.E.; Lorenzo, J.C.; Ojeda, E.; Quiñones, J.; Mora, N.; Sánchez, M.; Iglesias, A.; Martínez, J.; Castillo, R. (2006) Methodology for the cryopreservation of calli with embryogenic structures for the culture of sugarcane. *Biotecnología Aplicada* 23(4), 360-375. ISSN: 1027-2852

Martínez-Montero, M.E.; Martínez, J.; Castillo, R.; Engelmann, F.; González-Arnao, M.T.; (2005) Cryopreservation of pineapple (*Ananas comosus* (L.) Merr) apices and calluses. *Acta Hort. (ISHS)* 666, 127-131. (ISSN 0567-7572

Martinez-Montero, M.E.; Martinez, J.; Engelmann, F. (2008) Cryopreservation of sugarcane somatic embryos. *Cryo-Letters*, 29(3), 229-242. ISSN: 0143-2044

Martinez-Montero, M.E.; Mora, N.; Quiñones, J.; Gonzalez-Arnao, M.T.; Engelmann, F.; Lorenzo, J.C. (2002a) Effect of cryopreservation on the structural and functional integrity of cell membranes of sugarcane embryogenic calluses. *Cryo-Letters* 23, 237-244. ISSN: 0143-2044

Martinez-Montero, M.E.; Ojeda, E.; Espinosa, A.; Sánchez, M.; Castillo, R.; Gonzalez-Arnao, M.T.; Engelmann, F.; Lorenzo, J.C. (2002b) Field performance of cryopreserved callus-derived sugarcane plants. *Cryo-Letters* 23(1), 21-26. ISSN: 0143-2044

Matsumoto T, Sakai A, Takahashi C y Yamada K (1995) Cryopreservation of *in vitro*-grown apical meristems of wasabi (Wasabia japonica) by encapsulation-vitrification method. CryoLetters. 16: 189-196. ISSN: 0143-2044

Matsumoto, T.; Sakai, A.; Takahashi, C.; Yamada, K. (1995) Cryopreservation of *in vitro*-grown apical meristems of wasabi (*Wasabia japonica*) by encapsulation-vitrification method. *Cryo-Letters* 16, 189-196. ISSN: 0143-2044

McGann, L.E. & Walterson, M.L. (1987) Cryoprotection by dimehyl sulfoxide and dimethyl sulfone. *Cryobiology* 24(1), (February 1987), 11-16. ISSN: 0011-2240

Merkle, S.A.; Parrot, W.A.; Flinn, B.S. (1995) Morphogenicaspects of somatic embryogenesis. In: *In vitro embryogenesis in plants. Current Plant Science and Biotechnology in Agriculture,* Thorpe, T.A. Ed. Kluwer Academic Publishers; 20, 155-204. ISBN: 0-7923-3149-4, Dordrecht, Netherlands.

Mittler, R.; Vanderauwera, S.; Gollery, M.; Van Breusegem, F. (2004) Reactive oxygen gene network of plants. *Trends Plant Sci.* 9, 490–498. ISSN: 1360-1385

Mix-Wagner, G.; Conner, A.J.; Cross, R.J. (2000) Survival and recovery of asparagus shoot tips after cryopreservation using the "droplet" method. *New Zeal. J. Crop & Hort Sci.,* 28, 283-287. ISSN: 0114-0671

Murashige, T.; Skoog, F. (1962) A revised medium for rapid growth and bioassays with tobacco tissue cultures. *Physiol. Plant.* 15: 473-497. ISSN: 0031-9317

Nieves, N.; Martinez-Montero, M.E.; Castillo, R.; Blanco, M.A.; González-Olmedo, J.L. (2001) Effect of abscisic acid and jasmonic acid on partial desiccation of encapsulated somatic embryos of sugarcane. *Plant Cell Tissue and Organ Culture* 65, 15-21. ISSN: 0167-6857

Niu, D.K.; Wang, M.G.; Wang, Y.F. (1997) Plant cellular osmotica. *Acta Biotheoretica* 45(2), 161-169. ISSN: 0001-5342

Panis, B. & Lambardi, M. (2006) Status of cryopreservation technologies in plants (crops and forest trees). In: *The role of Biotechnology in Exploring and Protecting Agricultural Genetic Resources,* Ruane, J. & Sonnino, A. Eds. (61-78) FAO, ISBN 92-5-105480-0, Rome, Italy.

Panis, B. & Thinh, N.T. (2001) *Cryopreservation of Musa germplasm,* INIBAP Technical Guideline 5. Escalant, J.V. & Sharrock, S. Eds., International Network for the Improvement of Banana and Plantain, ISBN: 2-910810-45-3, Montpellier, France. IPGRI, Rome, Italy.

Panis, B.; Piette, B.; Swennen, R. (2005) Droplet vitrification of apical meristems: a cryopreservation protocol applicable to all *Musaceae. Plant Sci.* 168, 45–55. ISSN: 0168-9452

Parsegian, V.A.; Rand, R.P.; Rau, D.C. (1995) Macromolecules and water: Probing with osmotic stress. In: *Methods in Enzymology: Energetics of Biological Macromolecules.* Johnson, M.L. & Ackers, G.K. Eds. 259, (43-94), ISBN: 978-0-12-182160-9, Academic Press Inc., New York.

Parsegian, V.A.; Rand, R.P.; Rau, D.C. (2000) Osmotic stress, crowding, preferential hydration, and binding: A comparison of perspectives. *Proceedings of the National Academy of Sciences USA,* 3987-3992. ISSN-0027-8424

Paulet, F.; Engelmann, F.; Glaszmann, J.C. (1993) Cryopreservation of apices of in vitro plantlets of sugarcane (Saccharum sp. hybrids) using encapsulation-dehydration. Plant Cell Rep. 12, 525-529. ISSN: 0721-7714

Pena, M. & Stay, M. (1997) Ventajas de la propagacion in vitro. Canaveral 3(24), 473–497. ISSN: 1026-0781

Pérez, J.N.; Jiménez, E.; Gómez, R. (1999) Mejora genética de la caña de azúcar mediante la inducción de mutaciones y la selección in vitro. In: Biodiversidad y Biotecnología de la Caña de Azúcar. A Arencibia & MT Cornide (eds.). pp. 79-92, ISBN: 959-235-015-9, Elfos Scientiae, Cuba.

Rabe, E. and C. J. Lovatt (1984). De novo arginine biosynthesis in leaves of phosphorus-deficient Citrus and Poncirus species. Plant Physiology 76(3): 747-752. ISSN: 0032-0889

Rasmussen, P.H.; Jorgensen, B. & Nielsen, J. (1997) Cryoprotective properties of proline in cod muscle studied by differential scanning calorimetry. Cryo-Letters 18, 293-300. ISSN: 0143-2044

Reed, B.M. (2001) Implementing cryogenic storage of clonally propagated plants. Cryo Letters 22, 97-104. ISSN: 0143-2044

Reinhoud, P.J.; Van Iren, F.; Kijne, J.W. (2000) Cryopreservation of undifferentiated plant cells. In: Cryopreservation of tropical plant germplasm - Current research progress and applications. Engelmann, F.; Takagi, H., Eds. (91-102) Tsukuba: JIRCAS; Rome: IPGRI. ISBN 92-9043-428-7, Rome, Italy.

Ruan, K.; Xu, Ch.; Li, T.; Li, J.; Lange, R.; Balny, C. (2003) The thermodynamic analysis of protein stabilization by sucrose and glycerol against pressure-induced unfolding: The typical example of the 33-kDa protein from spinach photosystem II. European Journal of Biochemistry 270, 1654–1661. ISSN: 0014-2956

Rudolph, A.S. & Crowe, J.H. (1985) Membrane stabilization during freezing: the role of two natural cryoprotectants, trehalose and proline. Cryobiology 22, 367-377. ISSN: 0011-2240

Sakai, A. & Engelmann, F. (2007) Vitrification, encapsulation-vitrification and droplet-vitrification: a review. Cryo-Letters 28, 151–172. ISSN: 0143-2044

Sakai, A.; Kobayashi, S.; Oiyama, I. (1990) Cryopreservation of nucellar cells of navel orange (Citrus sinensis Osb. var brasiliensis Tanaka) by vitrification. Plant Cell Rep. 9, 30-33. ISSN: 0721-7714

Sakai, A.; Matsumoto, T.; Hirai, D.; Charoensub, R. (2002) Survival of tropical apices cooled to -196°C by vitrification. In: Plant cold hardiness, gene regulation and genetic engineering. Li, P.H.; Palva, E.T. Eds., 109-119, Kluwer Academic/Plenum Publishers, ISBN: 0306472864, New York.

Schäfer-Menuhr, A.; Schumacher, H.M.; Mix-Wagner, G. (1997) Long-term storage of old potato varieties by cryopreservation of shoot-tips in liquid nitrogen. Plant Genet. Res. Newsletter 111, 19-24. ISSN: 1020-3362

Semenova, M.G.; Antipova, A.S.; Belyakova, L.E. (2002) Food protein interactions in sugar solutions. Current Opinions in Colloid Interface Science 7, 438–444. ISSN: 1359-0294

Spiteller, G. (1996) Enzymic lipid peroxidation-α consequence of cell injury? Free Radical Biology and Medicine 21(7), 1003-1009. ISSN: 0891-5849

Stanley, C.; Krueger, S.; Parsegian, V.A.; Rau, D.C. (2008) Protein structure and hydration probed by SANS and osmotic stress. *Biophysical Journal* 94, 2777–2789, (April 2008), ISSN: 0006-3495

Sun, W.Q. (1999) State and phase transition behaviors of *Quercus rubra* seed axes and cotyledonary tissues: relevance to the desiccation sensitivity and cryopreservation of recalcitrant seeds. *Cryobiology*, 38(4), (June 1999), 372-385. ISSN: 0011-2240

Swan, T.W.; O'Hare, D.; Gill, R.A.; Lynch, P.T. (1999) Influence of preculture conditions on the post-thaw recovery of suspension cultures of Jerusalem artichoke (*Helianthus tuberosus* L.). *Cryo-Letters* 20, 325-336. ISSN: 0143-2044

Swartz, H.J. (1991) Post culture behaviour: genetic and epigenetic effects and related problems. In: *Micropropagation. Technology and Application.* PC Debergh & RH Zimmermann (eds.). pp. 95-121, Kluwer Academic Publishers, ISBN: 9780792308195, Dordrecht, Netherlands.

Tapia, R.; Castillo, R.; Blanco, M.A.; González, J.L.; Sánchez, M.; Rodríguez, Y. (1999) Induction, maduration and encapsulation of sugarcane somatic (*Saccharum* spp) cv CP5243. *Biotecnología Aplicada* 16, 20-23. ISSN: 1027-2852

Taylor, P.; Ko, H.; Adkins, S.; Rathus, C.; Birch, R. (1992) Establishment of embriogenic callus and high protoplast yielding suspension cultures of sugarcane (*Saccharum* spp. hybrids) *Plant Cell Tissue and Organ Culture* 28: 69-78. ISSN: 0167-6857

Tetteroo, F.A.A. (1996) *Desiccation tolerance of somatic embryoids.* PhD Thesis, Wageningen Agricultural University, The Nederlands. ISBN 90-5485-517-7

Thierry, C.; Florin, B.; Petiard, V. (1999) Changes in protein metabolism during the acquisition of tolerance to cryopreservation of carrot somatic embryos. *Plant Physiol. Biochem.* 37, 145-154. ISSN: 0981-9428

Thinh, N.T.; Takagi, H.; Yashima, S. (1999) Cryopreservation of in vitro-grown shoot tips of banana (*Musa* spp.) by vitrification method. *Cryo-Letters* 20(3), 163-174. ISSN: 0143-2044

Thomashow, M.F. (1999) Plant cold acclimation: Freezing tolerance genes and regulatory mechanisms. *Annu. Rev. Plant Physiol. Plant Mol. Biol* 50, 571-599. ISSN:1040-2519

Timasheff, S.N. (1993) The Control of Protein Stability and Association by Weak Interactions with Water: How Do Solvents Affect These Processes? *Annual Review of Biophysical and Biomolecular Structure* 22, 67-97, (June 1993), ISSN: 1056-8700

Turner, S.; Senaratna, T.; Touchell, D.; Bunn, E.; Dixon, K.; Tan, B. (2001) Stereochemical arrangement of hydroxyl groups in sugar and polyalcohol molecules as an important factor in effective cryopreservation. *Plant Science* 160, 489-497. ISSN: 0168-9452

Turner, S.R.; Krauss, S.L.; Bunn, E. Senaratna, T.; Dixon, K.W.; Tan, B.; Touchell, D.H. (2001) Genetic fidelity and viability of *Anigozanthos viridis* following tissue culture, cold storage and cryopreservation. *Plant Science* 161, 1099-1106. ISSN: 0168-9452

Tyler, N.; Stushnoff, C.; Gusta, L.V. (1988) Freezing of water in dormant vegetative apple buds in relation to cryopreservation. *Plant Physiol* 87, 201-205. ISSN: 0032-0889

Ulrich, J.M.; Finkle, B.J.; Moore, P.H. (1984) Frozen preservation of cultured sugarcane cells. *SugarCane* 3, 11-14.

Ulrich, J.M.; Finkle, B.J.; Moore, P.H.; Ginoza H. (1979) Effect of a mixture of cryoprotectants in attaining liquid nitrogen survival of callus culturesof a tropical plant. *Cryobiology* 16, 550-556. ISSN: 0011-2240

Verleysen, H.; Samyn, G.; Van Bockstaele, E.; Debergh, P. (2004) Evaluation of analytical techniques to predict viability fter cryopreservation *Plant Cell Tissue and Organ Culture* 77, 11-21. ISSN: 0167-6857

Volk, G.M. & Walters, C. (2006) Plant vitrification solution 2 lowers water content and alters freezing behaviour in shoot tips during cryoprotection. *Cryobiology* 52, 48–61. ISSN: 0011-2240

Watanabe, K.; Kuriyama, A.; Kawai, F.; Kanamori, M. (1999) Effect of cryoprotectant treatment and post-thaw washing on the survival of cultured rice (*Oryza sativa* L.) cells after cryopreservation. *Cryo-Letters* 20, 377-382. ISSN: 0143-2044

Withers, L.A. & King, P.J. (1980) A simple freezing unit and routine cryopreservation method for plant cell cultures. *Cryo-Letters* 1, 213-220. ISSN: 0143-2044

Withers, L.A. (1985) Cryopreservation of cultures plant cells and protoplasts. In: *Cryopreservation of plant cells and organs.* Kartha, K.K. Ed., 243-267, CRC Press, ISBN 0-8493-6102-8, Boca Raton. USA.

Yang, M.H. & Schaich, K.M. (1996) Factors affecting DNA damage caused by lipid hydroperoxides and aldehydes. *Free Radical Biology and Medicine* 20(2), (July 1995), 225-236. ISSN: 0891-5849

Yoshida, S.; Hattanda, Y.; Suyama, T. (1993) Variationsin chilling sensitivity of suspension cultured cells of mung bean (*Vigna radiata* (L.) Wilczek) during the growth cycle. *Plant Cell Physiology* 34, 673-679. ISSN 0032-0781

9

Comparison of Cryopreservation Methods of Vegetatively Propagated Crops Based on Thermal Analysis

Jiří Zámečník, Miloš Faltus, Alois Bilavčík
and Renata Kotková
Crop Research Institute
The Czech Republic

1. Introduction

There is a trend to preserve the plant germplasm by not only conventional *ex situ* methods or *in vitro* techniques, but also, more recently, by cryopreservation. Cryopreservation techniques are based on the storage of plant samples at very low temperature at which practically no chemical reactions occur and consequently, neither aging nor genetic changes of plant material. There has been a great development progress of cryopreservation methods during last years. Cryopreservation becomes a highly utilized technique for germplasm conservation. Generally the cryopreservation is storage of the samples. The samples can be e.g. organs and shoots tips from *in vitro* culture, or from the field, such as mature, immature bulbils, cloves of garlic or dormant buds of fruit trees, at the ultra-low temperature (mainly –196 °C, the temperature of liquid nitrogen).

Although the technique was introduced for plants in the '70s, it has never been applied on a wide scale due to the high cost of cryo-freezers; indeed, it was used in order to escape the formation of lethal intracellular ice crystals, time-consuming and laborious slow-cooling procedures. A new cryogenic - vitrification technique is now available, aiming at the direct immersion of plant specimens from tissue cultures in liquid nitrogen, without resorting to an expensive apparatus for slow cooling and with a considerable simplification of the procedures (Benson, 2008). The vitrification method simplifies cryogenic process and makes possible an increased application of cryopreservation on wide-range plant genetic resources. The glassy state is the objective status of cryopreservation methods named vitrification.

The aim of this study is a comparison of different cryopreservation methods based on the vitrification achieved by dehydration and glass transition temperature (T_g), and their efficiency towards optimal regeneration of vegetatively propagated plants. The thermal characteristics, evaluation of frozen water content, and the glass transition temperature were measured by a differential scanning calorimeter.

2. Importance of cryopreservation of vegetatively propagated plants

Some of vegetatively propagated plants are not able to reproduce by seeds e.g. garlic plant (*Allium sativum* L.). The only way how to propagate it is to use its cloves or bulbils for seeding plants for further growing. The vegetatively propagated plant germplasm is endangered by abiotic and biotic factors in the field conditions. Although the production area of many vegetatively propagated plants has been decreasing, many local cultivars and varieties remain. In the presence of decreasing cultivar variability in production areas, diminishing of old orchards, as well as appearance of diseases close to field collection areas, the question of safely maintaining the broad genetic potential of fruit trees is arising.

Two safe methods ensure vegetatively propagated plant germplasm maintenance with a low risk of loss: slow-growth *in vitro* culture and the cryopreservation methods. Advantages of *in vitro* collection are aseptic and stable conditions of the cultivation and availability of the material during the year. A disadvantage is the necessity of sequential plant multiplication. Advantages of cryo-collection are low costs for its long-term maintenance and material stability. Disadvantages are a longer time for the plant to recover from stored material and a rather high input costs of the cryopreservation procedure. The best way how to maintain germplasm is the combination of both methods. The base collection should be maintained by *in vitro* collection that provides the material in case of requirements. Core collection of the most valuable material, should be backed-up by cryo-collection for long-term storage, and plants are recovered just in case the genotype is lost from the base collection. For that reason, important vegetatively propagated plant collections have started to introduce accessions to slow-growth *in vitro* cultures and simultaneously in cryo-collection in liquid nitrogen (Gonzalez-Arnao *et al.*, 2008; Keller *et al.*, 2008; Kim *et al.*, 2006).

3. Cryopreservation methods

The latest results from the field of low temperature biology suggest that the main factor influencing the success of the cryopreservation method is the maintenance of a glassy state in plant samples and the avoidance of ice nucleation. The danger of ice nucleation and subsequent ice crystallization leading to frost damage during cooling and rewarming of samples is considered as a critical point of cryopreservation. That is the reason why many of the new progressive methods use and involve a glassy state in plant material intended for cryopreservation. Knowledge of the glass transition temperature is useful not only for improving methods involving glassy state in plant shoots tips. It also provides information essential for the long-term storage of shoot tips.

This biotechnology is based on the induction of the vitrification status – glass induction by dehydration, addition of cryoprotectants and a very fast decrease in temperature. Vitrification can be achieved in a number of ways (Sakai & Engelmann, 2007) but they usually all have the results of increased solute concentration to a critical viscosity. Low water content minimizes the ice crystallization that is potentially dangerous for plant cells and increases the temperature of glass transition. Supposing that the change of water status in the certain range is not limiting for plant regeneration. Plant Vitrification Solutions (PVS) marked with numbers according to the specific mixture of basic cryoprotectants and their concentrations are usually used for osmotic dehydration. Another cryopreservation method used, is based on desiccation in the air-flow cabinet. It is defined with the flow rate, temperature and humidity or on desiccation over various saturated salt solutions with steady-state activity of water.

3.1 Cryoprotectants involved in vitrification method

The cryopreservation method using a vitrification solution was first described by (Luyet, 1937). The vitrification solutions were firstly named, according to the first author of the publication and later the vitrification solutions have abbreviated names from Plant Vitrification Solution (PVS) with a number according to the time of their first appearance in the literature. The main ones are Luyet (1937), Fahy (1985), Steponkus (Langis & Steponkus, 1990), PVS1 (Uragami et al., 1989; Towill, 1990), PVS2 (Suzuki et al., 2008), PVS3, PVS4, PVS5 (Nishizawa et al., 1993), VS6 (Liu et al., 2004a), PVSL (Liu et al., 2004b) VSL (Suzuki et al., 2008), with different concentration and combination of the main four components: dimethylsulfoxide, sucrose, glycerol and ethylene glycol. The increased efficiency of vitrification methods was achieved by treating plants in the pre-cultivation step before cryopreservation of plant shoot tips in so called Loading Solution (LS) (Dumet et al., 2002; Matsumoto & Sakai, 1995; Sakai et al., 1991; Sakai & Engelmann, 2007). The cryoprotective substances should fulfil several basic parameters, such as cell permeability, viscosity, toxicity and the minimum concentration necessary for the vitrification, which eliminates the formation of ice crystals.

Cryoprotective substances help to ensure the stability of membranes and enzymes in the subsequent dehydration by vitrification solutions and to avoid the formation of ice crystals (Kartha & Leung, 1979; Kim et al., 2006). The samples are exposed to a several hour-long treatment by some cryoprotective substances, and then they are plunge-frozen in liquid nitrogen. The effect of cryoprotective solution composition for plant regeneration was studied in different plant species (Ellis et al., 2006; Kim et al., 2004; Kim et al., 2009; Tanaka et al., 2004).

In the most recent approaches to the garlic cryopreservation, vitrification method can be induced by treating the shoot tips of plantlets with a highly concentrated a mixture of glycerol and sucrose. (Nishizawa et al., 1993) developed Plant Vitrification Solution 3 (PVS3) with 50% glycerol (w/v) and 50% sucrose (w/v) in water. It is noteworthy that, following these procedures, the plant specimens can be directly plunged into liquid nitrogen, where they can be stored for an indefinite period of time without undergoing the risks of contamination or genetic alterations.

3.2 Methods based on dehydration

Potato (*Solanum tuberosum* L.) is a plant species sensitive to frost temperatures. Cryoprotocol for potato has to solve the problem of how to overcome temperature between 0 °C and –130 °C during cooling and warming without ice crystal growth and cell damage. Cold acclimation is not appropriate as pre-cultivation for potato plant (Hirai & Sakai, 1999; Schafer-Menuhr et al., 1996; Kaczmarczyk, 2008). The only method for potato vitrification is a water content decrease in samples, and than the rapid cooling and warming rate. Water content decrease is achieved by preculturing explants with osmotic compounds, air desiccation or vitrification. On bases, vitrification (Sarkar & Naik, 1998), droplet (Schafer-Menuhr et al., 1996) and recently vitrification-droplet (Halmagyi et al., 2004; Schafer-Menuhr et al., 1996) methods were developed or adapted for potato.

3.3 Encapsulation-dehydration

One of the other cryopreservation methods is encapsulation-dehydration. The shoot tips were encapsulated in an alginate gel. Experiments with dynamic dehydration studies demonstrated the necessity of meristems encapsulation (Benson et al., 1996; Grospietsch et *al.*, 1999; Hirai & Sakai, 1999). The encapsulation of shoot tips prolongs the dehydration period up to seven hours at a low relative humidity. The alginate beads without shoot tips had approximately the same dehydration-time curve. On the contrary no encapsulated shoot tips were completely dehydrated up to 1hour. The static dehydration of shoot tips was done over the various saturated salt solutions.

4. Cryoprotocols

4.1 Garlic

The unripe topsets were surface sterilized by chloramines and 75% ethanol and from this point, all preparations were performed under sterile conditions using sterile instruments and culture media in a laminar flow box. Opening the surface sterilized topset in sterile condition, all the inside structures were sterile (Fig. 1a). The sterile unripe bulblets were removed and cuts were made to the clusters or clumps of 3-8 bulbils. The bulbils varied in the thickness (approximately 2 mm) and in the length (3-5 mm) depending on the genotype and stage of ripening. Inside the unripe bulbils shoot tips were (Fig. 2a) with meristematic tissue.

Pre-culture of unripe clusters of bulbils was done on the MS culture medium (Murashige & Skoog, 1962) with 0,2 mg L^{-1} BAP and 0,02 NAA mg L^{-1} with 10 % sucrose for 20-24 h at 22 °C and 16 h light in Petri dishes sealed with Parafilm.

Fig. 1. Plants grown in *in vitro* conditions ready for dissection of shoot tips: (a) Garlic. Scale bar, 5 mm; (b) Potato. Scale bar, 1 mm; (c) Hop. Scale bar, 5 mm and (d) Apple tree. Scale bar, 10 mm.

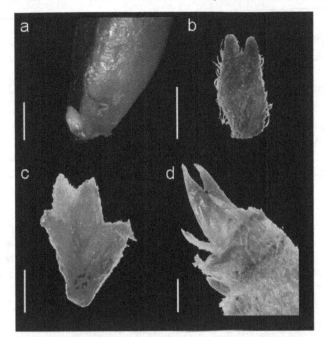

Fig. 2. The size and shape of shoot tips used for cryopreservation. (a) Garlic. Scale bar, 1 mm; (b) Potato. Scale bar, 0,25mm; (c) Hop. Scale bar, 0,25 mm and (d) Apple tree. Scale bar, 1 mm.

Cryoprotocol Steps	The Procedure
Unripe bulbils dehydration	
	Immersion in the loading solution (13,7 % (w/v) sucrose + 18,4 % (w/v) glycerol in the liquid medium) (Sakai *et al.*, 1991) for 20 minutes
Cryopreservation	
	Dehydration by PVS3 (Nishizawa *et al.*, 1993) at a laboratory temperature for 2 hours
	Removing and adding fresh PVS3 before freezing
	Aluminum foil stripes with 5-10 clusters of unripe bulbils plunged directly into liquid nitrogen at least for one hour in liquid nitrogen (Sarkar & Naik, 1998)
Thawing	
	Rapid warming immersion into a 40 °C water bath for 30-120 seconds for thawing
Survival and regeneration evaluation	
	Sub-culture on MS medium supplemented with 0,2 mg L^{-1}BAP and 0,02 NAA mg L^{-1} with 3 % sucrose for seven days in the dark
	Evaluation of survival after two weeks (Fig. 12a.)
	Evaluation of regeneration after 8-10 weeks

Table 1. Cryopreservation steps of garlic

4.2 Potato

Potato explants (Fig. 1b.) were multiplied by nodal cuttings in plastic boxes (Vitro Vent container, Duchefa) on 100 ml modified MS medium (Grospietsch et al., 1999) with 7 g L⁻¹ agar and 30 g L⁻¹ sucrose, without myo-inositol and phytohormones, with a decreased amount of nitrogen at pH 5,5. Nodal cuttings were cultivated at 22 ± 1 °C, 80 μmol m⁻² s⁻¹ and photoperiod 16/8 h light/dark (L/D) (Fig. 1b). Subculture interval was 3-4 weeks.

Nodal cuttings were planted in the same conditions as the pre-cultured plants but only 50 ml medium was used per one box. After 4 day pre-culture, lateral buds elongated to at least 1 mm (Fig. 2b). Subsequently, 25 ml 2 M sucrose was added into each container and explants were cultivated at the same conditions for the next 5-6 days.

Cryoprotocol Steps	The Procedure
Shoot tips dehydration	
	Shoot tips (Fig. 2b.) (1-2 mm) were transferred onto a filter paper moistened with 14 ml 0,7 M sucrose and phytohormones of the same composition as in a recovery medium (0,5 mg L⁻¹ IAA + 0,5 mg L⁻¹ Kinetin + 0.2 mg L⁻¹ GA3) at 22 ± 1 °C and photoperiod 16/8 h (L/D) and shielded with a leaf of paper for overnight. The second day the shoot tips were transferred (20 tips per foil) onto aluminum foils (20 x 6 x 0.05 mm) Dehydration above silicagel for 1,75 – 2 h.
Cryopreservation	
	Plunging aluminum foils into liquid nitrogen Stored in cryovials (two foils with 20 shoot tips per vial).
Thawing	
	Alluminum foils plunged rapidly into the water bath at a laboratory temperature. Transfer immediately onto the recovery medium (Grospietsch et al., 1999) with the same composition as the pre-culture medium but with phytohormones (0,5 mg L⁻¹ IAA + 0,.5 mg L⁻¹ Kinetin + 0,2 mg L⁻¹ GA3)
Survival and regeneration evaluation	
	Survival was defined by shoot tip growth and by green color and recovery as a new explant development (Fig. 12b.) Plant regeneration was evaluated 2 and 8 weeks after cryopreservation.

Table 2. Cryopreservation steps of potato

4.3 Hop

Maternal plants were cultivated on a multiplication medium without phytohormones at 22 ± 1 °C, 80 μmol m⁻² s⁻¹, photoperiod 16/8 h (L/D); subculture interval was 8 weeks. Modified solid medium (Murashige & Skoog, 1962) without casein and myoinositol, with decreased amount of nitrogen (25 % (w/v) of NH_4NO_3 and 50 % (w/v) of KNO_3 of the original Murashige and Skoog medium), with 40 g L⁻¹ glucose, pH 5,5 without phytohormones was used as the multiplication medium (MSH).

Nodal cuttings were planted in the same conditions as during explant multiplication but only 50 ml medium was used per one box (Fig 1c). After 7-10 d pre-culture, lateral buds elongated to 1-2 mm. Then the explants were transferred into cold acclimation conditions at 4 °C for 7-10 days. Subsequently 25 ml 0,7 M sucrose was added into each container and explants (Fig. 1c.) were cultivated at the same conditions for the next 7-10 days.

Cryopreservation steps for hop shoot tips were the same as for potato (see Tab. 2).

4.4 Apple tree

In vitro plants (Fig. 1d) were cultivated in 100 ml Ehrlenmeyer flasks with 20 ml of MS medium, 3% (w/v) sucrose, 6 g L^{-1} agar, supplemented with GA3 1 mg L^{-1}, BAP 1 mg L^{-1}, IBA 1 mg L^{-1}, at 20 ± 1 °C, 8/16 (L/D) photoperiod of light intensity 100 μmol m^{-2} s^{-1}.

Cryoprotocol Steps	The Procedure
Shoot tips dehydration	
	One to two week subcultivation on fresh MS medium,
	Cold hardening for 4-6 weeks at +4± 1 °C, short photoperiod (8/16, L/D) of light intensity 25 μmol m^{-2} s^{-1}
	Petri dish with 16 ml of MS medium and poured with 16 ml of 2M sucrose solution for 48 hours
	Encapsulation of shoot tips (Fig. 2d.) with meristematic cells in 3% w/v dipped into alginate in 0,75M sucrose for 10 min and then drop into 0,1 M CaCl2 in 0,6 M sucrose for 10 min to net the alginate and form a bead
Cryopreservation	
	Beads are gently dried on a sterile filter paper
	Additional dehydration in the laminar flow box at laboratory temperature and dehydrated for different time up to 4 -5 hours, Placed in 2 ml cryovials and plunged in liquid nitrogen.
Thawing	
	Shoot tips with beads were warmed up by plunging cryovials in 40 °C water
Survival and regeneration evaluation	
	Survival was defined by shoot tip growth and by green color and regenerated as a new explant development (Fig. 12d.)

Table 3. Cryopreservation steps of apple tree

5. Thermal analysis

Differential scanning calorimetry (DSC) belongs to thermal methods that can be used for measurement and determination of phase and glass transitions for cryopreservation. In principle, the DSC measures the temperatures and heat flows associated with transitions in plant material as a function of time and temperature. It gives information about endothermic or exothermic changes or changes in heat capacity. The obtained data can be used for determination of glass transition, temperature of ice nucleation, melting, boiling, crystallization time and kinetic reaction – the most important characteristics useful for cryopreservation (Zámečník & Faltus, 2009). The danger of nucleation and subsequent intracellular ice crystallization leading to frost damage during cooling and rewarming of the samples is considered a critical point of plant survival at ultra-low temperatures.

The differential scanning calorimetry method is based on the regulated decrease/increase temperature of the sample and reference and the measurement of temperature and heat flow corresponding to the sample. There are two different types of the differential scanning calorimeters. The power compensation DSC type directly measures heat release/uptake from the sample and the heat flow type measures differences of temperature between reference and sample and recalculates the differential heat flux. The most common cooling/heating rate of the sample is 10 °C min-1.

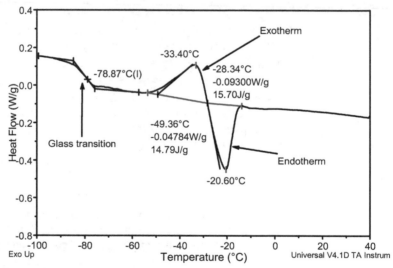

Fig. 3. Example of apple tree shoot tips heat flow response to the temperature. Cooling and warming rate was 10 °C min -1. Exo up.

In our experiments we used shoot tips of *in vitro* cultures of apple tree. Samples with different water content were obtained by air dehydration of alginate encapsulated shoot tips in the flow box or by dehydration at 4 °C of *in vitro* cultures. For DSC measurement dissected shoot tips were placed in aluminum sample pans and measured by Differential Scanning Calorimeter TA2920. Samples were cooled down to -120 °C (rate of 10 °C min-1). The data were collected during heating to 20 °C (rate of 10 °C min-1). The purge gas was either nitrogen or helium.

A Differential Scanning Calorimeter is used as a main tool in cryobiology to assist cryopreservation protocol development, to store thermograms as a documentation of cryo-protocols in the use and to keep information about stored samples and their thermal properties before, during and after cryopreservation (Benson *et al.*, 1996; Faltus & Zámečník, 2009; Šesták & Zámečník, 2007; Zámečník *et al.*, 2007)

There is an example (Fig. 3) of measured thermal characteristics of shoot tips of apple tree *in vitro* culture cv. Greensleeves by the DSC. Samples of an approximate weight of 10 mg were crimped in an aluminium sample pan and cooled from room temperature to -120 °C. Cooling and heating rate was 10 °C min[-1]. The glass transition, exothermic and endothermic characteristics were analyzed in detail during heating. Thermal characteristics were measured by DSC TA 2920 (TA Instruments) and evaluated by Universal Analysis 2000 for Windows (TA Instruments).

6. Water content and glass transition

6.1 Garlic

Water content during dehydration of garlic cv. Djambul 2 clusters of shoot tips by PVS3 (Fig. 4). Total amount of water (solid line) and amount of crystallized water (dash line) in shoot tips treated with PVS3 rapidly descend during the first 1,5 hours and further is constant. Crystallized water reaches minimum after 1,75 hours of PVS3 treatment. In this case the decrease of water content in the unripe bulbils is probably so low that it can have no further influence on the glass transition change. In comparison with the measurements in this study on the apple tree shoot tips (see below), the glass transition temperature increases with decreases of water after dehydration.

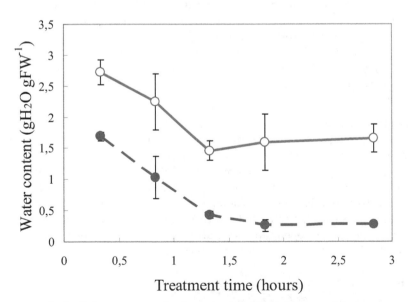

Fig. 4. Unripe garlic bulbils water content (empty circle) and the part of frozen water (full circles) during PVS3 treatment. Note: The unripe bulbils were in the loading solution first 20 minutes than they were immersed in to the PVS3. The bars are standard deviation of mean.

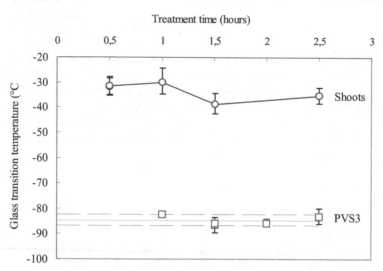

Fig. 5. Glass transition temperature of *Allium* shoot tips after moisture loss by dehydration in the Plant Vitrification Solution 3 (PVS3). Circles show the glass transition midpoint in shoot tips and bars show the onset and endset of glass transition. Squares show the glass transition of PVS3 in shoots. The full line is for the glass transition of PVS3 without shoots. The dashed line above is the endset of glass transition and below the onset of glass transition for PVS3.

Glass transition of garlic shoot tips was measured after different times of treatment – unripe bulbils in PVS3 at 23 °C (Fig. 5). At each curve, there were two S-shape heat flow changes during warming of the samples, typical for glass transition. The lower glass transition temperature on unripe bulbils heat flow curves coincides within the range of onset and endset of the glass transition temperature of PVS3 measured after unloading unripe bulbils. This glass transition temperature can be of PVS3 coating on the surface of the shoot tips immersed in PVS3.

The high glass transition temperature corresponds to glass transition of the shoot tips because at this range of temperature there were no thermal events on the PVS3 temperature dependent curve. There is no significant difference in the change of shoot tip glass transition changes from 0,5 to 2,5 hours of PVS3 treatment. The detectable glass transition was found between -30 °C and -39 °C. The average glass transition temperature is -33,5 °C after 0,5 hour. From these results it is obvious that the glass transition at higher temperature is for shoot tips saturated with PVS3. So, for the survival of shoot tips after thawing from liquid nitrogen, the second glass transition which occurs at higher temperatures is important (Zamecnik *et al.*, 2011).

6.2 Potato

Nodal cuttings were pre-cultured on medium with added sucrose solution. The final sucrose concentration in medium was 0,7 M. The importance of sucrose pre-treatment before potato cryopreservation proved by Grospietsch *et al.*, (1999) and Halmagyi *et al.*,

(2004). Halmagyi *et al.*, (2004) showed the highest plant regeneration after cryopreservation following a pre-treatment with 0,5 M sucrose. Similarly, Sarkar and Naik (1998) found a slightly negative effect of 0,7 M sucrose pre-treatment compared in comparison with 0,5 M or 0,3 M sucrose pre-treatment. In the present study the injury of potato explants was not observed after 0,75 M sucrose treatment.

Total water content in the shoot tips after nodal cutting pre-culture was approximately 5 g of H_2O per 1 g of dry mass (gH_2O g DW^{-1}) (Fig. 6.). Frozen water content in shoot tips was 4,3 gH_2O g DW^{-1} and the unfrozen 0,7 gH_2O g DW^{-1}. Subsequently shoot tips were isolated and loaded with 0,7 M sucrose in a Petri dish on filter paper for overnight. Total water content of shoot tips decreased to 2,1 gH_2O g DW^{-1}, from which 1,4 gH_2O g DW^{-1} represents the frozen water fraction and 0,7 gH_2O g DW^{-1} the unfrozen water fraction. Because the total water content and frozen water fraction decreased but the unfrozen fraction did not change, the ratio of frozen/unfrozen water content (WC_f/WC_u) decreased from 6 to 1,9. The following air dehydration resulted in a decrease of total water content due to both water fractions decrease. After 1,5h air dehydration above silicagel the total water content in shoot tips was 0,49 gH_2O g DW^{-1} from which the frozen water content was 0,09 gH_2O g DW^{-1}, and the unfrozen 0,4 gH_2O g DW^{-1}. Resulting WCf/WCu ratio decreased to 0,22. The prolonged dehydration decreased both water fractions. After 2h air dehydration above silicagel the total water content in shoot tips was 0,28 gH_2O g DW^{-1} from which 0,006 gH_2O g DW^{-1} belonged to the frozen fraction and 0,276 gH_2O g DW^{-1} to the unfrozen fraction. The WC_f/WC_u ratio decreased to 0,12 after 2h air dehydration of shoot tips above silicagel, which represents 2 % crystallized water of the total water content.

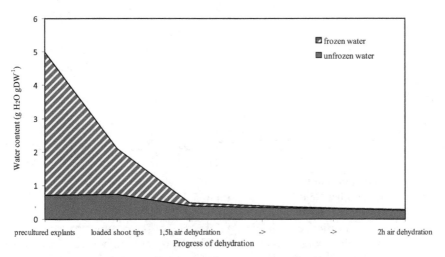

Fig. 6. The progress of dehydration of potato explants (cv. Désirée) after specific steps of cryoprotocol. Explants were pre-cultured on medium with 0,7 M sucrose. The isolated shoot tips were loaded with 0,7 M sucrose solution for overnight. The loaded shoot tips were dehydrated by dry air above silicagel for 2 hours. The amount of frozen and unfrozen water was determined by the DSC analysis.

The decrease in percentage of crystallized water in shoot tips during 1,75 to 2h air dehydration is illustrated in Fig. 6. The crystallized water content decreased from approximately 9% to 2%. Dehydration of shoot tips was connected to the glass transition temperature increase from -38 to -32 °C. The optimal water content of potato shoot tips was approximately 0,4 gH_2O g DW^{-1} that was obtained between 1,5h and 2h air dehydration above silicagel according to the size of particular genotype shoot tips. The temperature of glass transition was approximately -35 °C and the amount of frozen water was very small but still detectable (Fig. 7). Decrease in water content and onset of melting temperature was also found after dehydration by PVS2 solution or 10 % DMSO (Kaczmarczyk, 2008, Kaczmarczyk *et al.* 2011). However the T_g found by these cryoprotectants was lower than -100 °C. The higher temperature of glass transition found in this study indicated a higher stability of material stored at ultra-low temperatures.

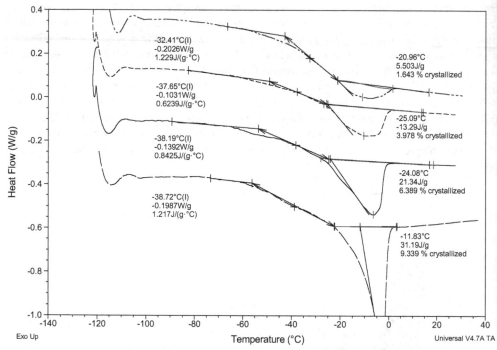

Fig. 7. DSC curves of air dehydrated potato shoot tips (cv. Désirée) air-dehydrated above silicagel for 1,75 to 2 hours. Heat flow was evaluated during warming the samples from -130 to 30 °C by ramp temperature 10 °C min^{-1}. Glass transitions were defined by the temperature of glass transition, change of heat flow per g of sample and change in specific heat capacity (C_p). Melting exotherms are defined by the onset temperature of melting, enthalpy change of thermal event, and crystallinity of water. Curves are shifted along y-axis for clarity according to crystallized water.

The most valuable accessions from sub-collection of old potato cultivars of the Czech origin were selected from the potato in vitro-bank at the Crop Research Institute (CRI) to store them by cryopreservation method. A new cryopreservation method based on nodal cutting osmotic pre-treatment, shoot tips sucrose loading and their air dehydration on aluminum foils was used for storage of 58 selected potato. All plant accessions prepared for storage in cryo-bank were virus-free. Average post-thaw recovery of hop and potato was 36 % and 25 %, respectively. Recovery of new plants was successful in all tested genotypes.

6.3 Hop

Isolated hop shoot tips (cv. Saazer) were dehydrated by air above silicagel (Fig. 8). Water content was 2,4 g water per 1g dry mass before air dehydration. The highest water decrease was measured during the first 30 minutes of dehydration. Water content of hop shoot tips was 0,68 gH_2O gDW^{-1} after 32 minutes of dehydration. Shoot tips water content decreased below 0,5 gH_2O gDW^{-1} after 70 minutes of dehydration and reached 0,4 gH_2O gDW^{-1} after 100 minutes of dehydration. After 120 minutes the shoot tips water content was 0,37 gH_2O g DW^{-1}. The plant regeneration depended on the time of dehydration, which was influenced by the shoot tips water content. The highest explant regeneration was achieved after 90 minutes of dehydration at a water content close to 0,4 gH_2O g DW^{-1}.

In a former study, a decrease in the endothermic peak was found during air dehydration by encapsulation-dehydration method used for hop cryopreservation (Martinez et al.,1998; Martinez et al., 1999; Martinez & Revilla, 1998). A negligible amount of freezable water was detected in shoot tips after the water content decreased to 18 % and no freezable water was found at a water content of 14 %. The glass transition temperature was found at a water content of 18 % and lower. The temperature of glass transition increased with a decrease of water content (Fig. 9).

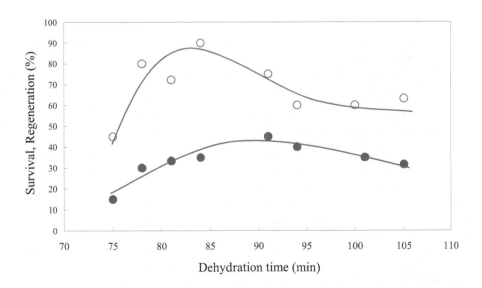

Fig. 8. Survival (empty circles) and regeneration (full circles) of hop explants during dehydration (cv. Saazer).

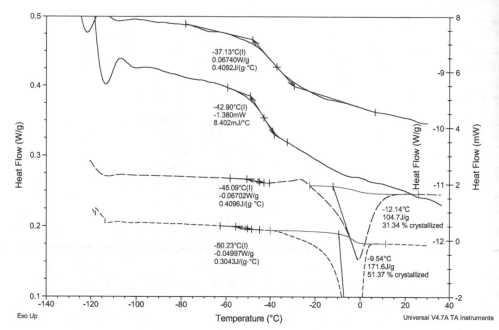

Fig. 9. Heat flow response to the temperature of hop shoot tips. Glass transition temperature of hop shoot tips after moisture loss by dehydration above silicagel. Cooling and warming rate was 10 °C min^{-1}. Curves are shifted along y-axis for clarity.

6.4 Apple tree

In Fig. 10, there is an example of measured thermal characteristics of encapsulated shoot tips of apple tree *in vitro* culture cv. Greensleeves by the DSC. Samples of approximate weight of 10 mg were crimped in an aluminium sample pan and cooled from room temperature to -120 °C. The cooling and heating rate was 10 °C min^{-1}. The glass transition, exotherm and endotherm characteristics were analysed in detail during heating. Thermal characteristics were measured by DSC TA 2920 (TA Instruments) and evaluated by Universal Analysis 2000 for Windows (TA Instruments).

The course of dehydration of encapsulated shoot tips of *in vitro* cultures of apple tree cv. Greensleeves in an open Petri dish exposed to air flow in laminar flow hood at laboratory temperature is demonstrated in Fig. 10. The determination of water content of 20 encapsulated shoot tips placed in the Petri dish was done by weighing the shoot tips during dehydration after approximately 4 hours of drying to the constant weight in an oven (105 °C). The water content was calculated as a proportion of g of water to g of dry matter. From measurement it was evident, that the dehydration consists of two parts; a faster one at the beginning and a slower one after approximately 2 hours of dehydration. From 2 h of dehydration, the level of water content in encapsulated apple tree shoot tips is almost constant. In Fig. 10, there are thermal characteristics (during heating) of encapsulated apple tree shoot tips during their dehydration in the air flow in flow hood. The more dehydrated the samples, the higher the glass transition temperature (characterised by the inflex point I) was measured and no endothermal events representing water in ice crystal form were detected below 0,4 gH_2O g DW^{-1}. The value of 0,4 gH_2O g DW^{-1} dehydration level corresponds to the levels recommended also by other authors (Gupta & Reed, 2006; Martinez *et al.*, 1999; Wu *et al.*, 1999). The integration of endotherm areas of shoot tip and alginate confirms the importance of dehydration to the levels when ice crystals are not present in shoot tip tissues (Figs. 10,11). The energy counted as integration of the endothermic peak corresponded to the amount of frozen water; the less energy, the smaller amount of ice crystals in the sample. These thermal results led us to dehydrating encapsulated shoot tips below 0,4 gH_2O gDW^{-1}.

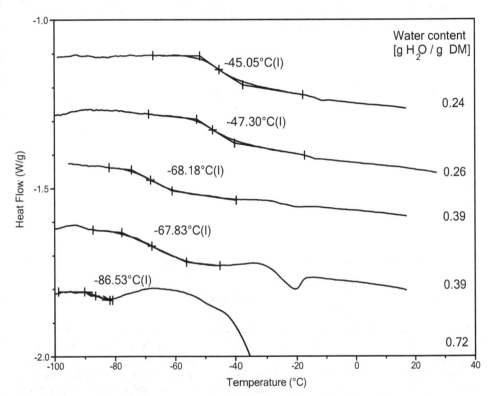

Fig. 10. Glass transition temperature as an inflection point of heat flow change of encapsulated apple tree shoot tips after water loss expressed as final water content (figures behind the end of the separate curves) by dehydration in the air flow. Curves are shifted along y-axis for clarity.

Dehydration curves corresponding to the loss of water from encapsulated *in vitro* shoot tips were measured (Figs. 10,11). During the dehydration procedure of cryopreservation The proper time/level of dehydration must be taken into consideration for successful cryopreservation.

The less water in plant tissues, the less probable damage from ice crystal formation and growth. On the other side plant tissues withstand only certain dehydration. The most appropriate level of dehydration is determined by DSC by measurement of frozen and unfrozen water (generally it is possible to say water in glassy state) (Fig. 11). After 4h dehydration of encapsulated *in vitro* shoot tips the water content decreased below 0,3 gH$_2$O gDW^{-1}. Water content decrease slows down markedly at the level of 0,6 gH$_2$O gDW^{-1}. From this level of dehydration, both exotherms and endotherms start to disappear which corresponded to the end of ice crystal formation and the start of glass transitions with high change of heat capacity (Fig. 11). The survival and regeneration of cryopreserved apple tree shoot tips, cultivar Greensleeves, were 75 % and 53 % respectively after 4h dehydration. Non-dehydrated shoot tips neither survived nor regenerated. Dehydration of shoot tips to the level of glass formation is a crucial factor for their survival at ultralow temperatures.

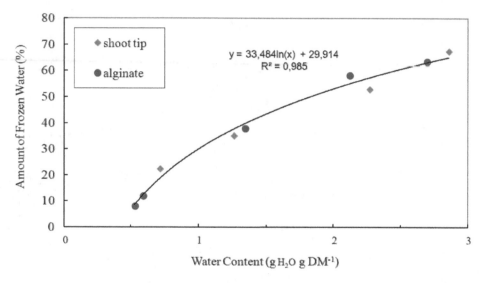

Fig. 11. The amount of frozen water of apple tree shoot tips and alginate beads during dehydration was as the integration of endotherm areas of shoot tip and alginate. Samples of an approximate weight of 3-10 mg were crimped in an aluminum sample pan and cooled from room temperature to -120 °C. The cooling and heating rate was 10 °C min^{-1}. The enthalpy counted as an integration of the endothermic peak corresponded to the amount of frozen water; the less enthalpy the smaller amount of ice crystals in the sample.

7. Regeneration of plants after cryopreservation

The regeneration rate of unripe garlic bulbils was close to 100 % in comparison with the lower regeneration rate of ripe bulbils. The results for ripe bulbils were done on 173 accessions (the measurements on ripe bulbils were not presented, Grospietsch unpublished). The average regeneration rate of ripe bulbils was 40 % and unripe bulbils near 100%. The optimized droplet-vitrification protocol was successfully applied to bulbil primordia of garlic varieties also with high regeneration percentages ranging between 77,4-95% (Engelmann, 2011; Kim *et al.*, 2006)

Fig. 12. Plants regenerated into new plants after immersion in and thawing from the liquid nitrogen: (a) Garlic. Scale bar, 5 mm; (b) Potato. Scale bar, 2 mm; (c) Hop. Scale bar, 1 mm and (d) Apple tree. Scale bar, 5 mm.

The average recovery after cryopreservation of fifty potato cultivars was 24,8% and average hop recovery was 30,5%. Plant recovery eas improved due to the cryoprotocol and media modifications and the average recovery of potato and hop in the year 2007 was 29,1% and 35,5%, respectively (Fig. 13). The highest frequency of plant recovery was near to the average recovery in both crops.

To improve the stability and safety of potato collection, the cryo-collection of the Czech potato germplasm was established. The sub-collection of old potato varieties of the Czech origin was selected as the most important part of potato germplasm kept in the Czech In Vitro Bank of Potato. Fifty eight selected genotypes were cryopreserved by a new method based on osmotic adjustment of explants with sucrose and following air-dehydration. Currently these 58 old potato cultivars of the Czech origin were backed up in cryo-collection at the CRI in Prague.

The differences in plant survival and regeneration exist either among species or cultivars. Example of survival and regeneration of apple tree *in vitro* cultures cryopreserved by the encapsulation-dehydration method are shown in (Tab. 4). The differences can be caused either by different reaction of cultivars in cryopreservation protocol or cold hardening conditions or *in vitro* cultivation. For example, the cultivar McIntosh belongs to very cold resistant cultivars and also has very high regeneration after cryopreservation. On the contrary, the very cold tender cultivar Zvonkové had no survival in laboratory frost tolerance test on dormant buds (data not shown) but *in vitro* cultures were able to survive the cryopreservation procedure, although with very weak regeneration.

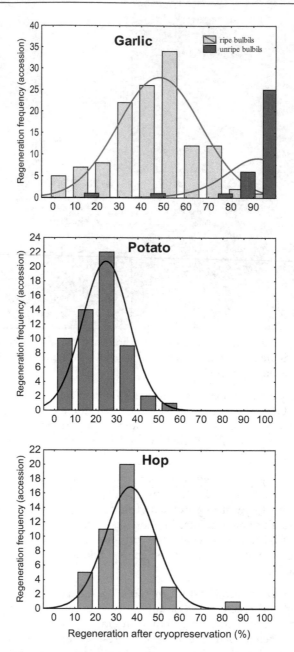

Fig. 13. Regeneration frequency of genotypes (accessions) of garlic, potato and hop stored in the Cryobank of vegetatively propagated crops. Altogether, 129 accessions were evaluated in ripe garlic bulbils, 34 accessions in unripe garlic bulbils, 58 accessions in potato and 50 in hop.

There is clear evidence for the necessity of physiological and biochemical adaptations of cryopreservation procedures according to the different demands of used cultivars to fulfill the needs for successful cryopreservation of apple tree *in vitro* germplasm.

Apple tree cultivar	Survival [%]	SD	Regrowth [%]	SD	n	Number of freezing
Alkmene	6 [a]	0,4	6 [a]	0,4	32	2
Golden Delicious	75 [bc]	27,9	55 [bcd]	24,7	39	3
Greensleeves	75 [bc]	15,0	53 [bcd]	7,5	30	2
Chodské	63 [bc]	17,6	46 [abcd]	13,8	84	5
Idared	34 [abc]	13,5	34 [abc]	13,5	45	2
Jonagold	63 [bc]	13,4	44 [abcd]	26,2	50	3
McIntosh	85 [c]	15,0	85 [d]	15,0	20	2
Prima	28 [ab]	15,0	21 [ab]	8,0	20	2
Rubín	78 [bc]	11,0	78 [cd]	11,0	32	2
Zvonkové	44 [abc]	18,7	4 [a]	5,4	56	4
Average	75	21,7	43	21,2	41	3

[a-d] average followed by the same index did not significantly differ at $P<0,05$ (analysis of variance – Duncan's test) SD - standard deviation ($P < 0,05$).

Table 4. Survival and regrowth of apple tree cultivars after encapsulation–dehydration cryopreservation protocol. (n = total number of shoot tips used for evaluations of regeneration).

In vitro shoot tips of apple tree were cryopreserved also by vitrification (PVS2) and pre-culture dehydration methods, but the results were not adequate (Tab. 5). Thus the basic approach of cryopreservation of fruit trees in our laboratory was the encapsulation-dehydration method. The average survival and regeneration of evaluated apple tree cultivars were 75 % and 53 %, respectively (Tab. 4). Similar values of regeneration were obtained by (Condello et al., 2011) with the droplet-vitrification method in two apple tree cultivars. On the other hand, (Wu et al., 2001) reached regeneration of up to 86 %. They recommended a prolonged subcultivation of mother plants and their cold acclimation, which decreased water content in shoot tips and subsequently increased the regeneration in all cryopreservation procedures they evaluated. According to our unpublished data and in concordance with other authors, (Wu et al., 2001; Condello et al., 2011), the adaptation of the *in vitro* mother plants appears to be one of the important steps for improving regeneration of cryopreserved cultivars with lower survival. The encapsulation-dehydration cryopreservation method is a suitable tool to conserve a broader spectrum of apple tree germplasm.

8. Comparison of different cryopreservation methods

The bases for achieving high regeneration after plant cryopreservation using a vitrification method needs to observe the time fulfilment in dehydrating procedures and prevention of tissue damage by chemical toxicity of the cryoprotectants. Otherwise, plant parts in solutes can be injured by the cryoprotectant, by strong osmotic stress during the cryoprotective solution treatment. The time for a high regeneration rate of plants after cryopreservation must be optimized. During this time the explants are dipped in the vitrification solutions, at the optimal temperature, which is involved in the procedure (Condello *et al.*, 2011; Faltus & Zamecnik, 2009; Sakai & Engelmann, 2007; Zámečník *et al.*, 2007).

The temperature-induced glasses-the point at which this occurs, is called the glass transition temperature (T_g) - molecular motion nearly ceases and the liquid becomes a glassy solid. Vitrification of cells and tissues is a physical process which avoids intracellular ice crystallization during ultra-rapid freezing by the transition of the aqueous solution of the symplast into an amorphous glassy state of the cells. As a consequence of the vitrification process, plant tissues are protected from the damage and remain viable during their long-term storage at –196 °C.

Experimental determination of glassy state in plants is complicated by the endothermic reaction overlapping with the glass transition. Ice crystallization as the first-order reaction has discontinuous change in heat capacity contrary to glass transition, which is characterized by heat capacity change. The main problem is to distinguish the endothermic reaction and the glass transition temperature during the measurement of the thermal events. Thermal analysis methods of glass transition temperature and temperature of ice crystals melting in plant tissues were determined. Standard Differential Scanning Calorimetry (DSC) method and Temperature Modulated Differential Scanning Calorimetry (TMDSC) were usually used (Condello *et al.*, 2011; Zámečník & Faltus, 2009).

Plant	Cryopreservation Method		
	Droplet -vitrification (PVS3)	Ultra-rapid freezing	Encapsulation -dehydration
Garlic	**	NT	*
Potato	*	**	*
Hop	NT	**	NT
Apple tree	*(§)	*	**

Table 5. Cryopreservation methods used in this study. Results obtained by cryopreservation methods with a high regeneration rate were presented only: ** - high regeneration rate; *- tested; NT- not tested; (§)-PVS2 was used.

The vitrification method, based on involvement of biological glass, requires a highly concentrated solution of cryoprotectant (from 5 to 8 M), at which the cells are osmotically dehydrated to a certain level. This level of dehydration is characterized by no frost-heaving of water and with a minimum, or no production of ice crystals. The cells treated by cryoprotectant are then vitrified before or during immersion into liquid nitrogen (Sakai *et al.*, 1991). Vitrification belongs to new well-developed procedures supplying frost dehydration of cells pursued at a low temperature. Namely, it removes most of the freezable water through the exposure of the plant shoot tips to a highly concentrated vitrification solution at temperatures above the freezing point.

Several procedures of cryopreservation of fruit trees germplasm, especially apple tree were evaluated, including two-step cryopreservation with controlled ice nucleation (Niino *et al.*, 1992; Niino & Sakai, 1992; Seufferheld *et al.*, 1999; Zhao *et al.*, 1999), and the type of vitrification method of extirpated *in vitro* cultures. The procedures were adjusted and corrected according to the thermal methods for determination of ice nucleation (Tyler *et al.*, 1988), glass transitions in plant material, and exothermic and endothermic characteristics of plant buds. A combination of encapsulation-dehydration cryoprotocol was chosen as the most appropriate system of *in vitro* cultures (Chang & Reed, 2001). The reaction of selected cultivars was different on *in vitro* sub-cultivation and subsequent cryopreservation protocol.

Many plants from the temperate and tropical region were successfully cryopreserved by the encapsulation-dehydration method, which belongs to cryopreservation methods. The ecapsulation-dehydration cryopreservation procedure is based on encapsulating shoot tips of pretreated *in vitro* plants with subsequent dehydration either in sterile air flow (Benson *et al.*, 1996; Dereuddre *et al.*, 1991) or above silicagel (Grospietsch *et al.*, 1999). Dehydrated beads with encapsulated shoot tips are in most cases plunged directly into liquid nitrogen or slowly frozen in programmable freezers (Fabre & Dereuddre, 1990); Zhao *et al.*, 2001). Rewarming proceeds either slowly by placing beads on Petri dishes or the cryotubes are placed in a water bath of temperature ranging from 25 to 45 °C for several minutes (Gupta & Reed, 2006; Matsumoto & Sakai, 1995). Survival and regeneration of shoot tips are evaluated after placing the re-warmed beads on cultivation medium which can be compound modified by phytohormones (their combination and concentration), to stimulate proliferation or eliminate the phenolic compounds (Paulet *et al.*, 1993). The medium can be softer in some cases to allow an easier regeneration from beads (Reed *et al.*, 2006) or even the explants can be extracted from the beads (Niino & Sakai, 1992). The importance of direct explant regeneration to the new plants without callus inter phase is important for the genetic stability after usage of this method.

9. Conclusion

The four main crops (garlic, potato, hop and apple tree) have been cryopreserved in the Czech Plant Cryobank. The methods of cryopreservation are based on cryoprotocol of the vitrification procedures (encapsulation-dehydration, dehydration by vitrification solution and a modified ultra–rapid freezing method based on preconditioning of the plant shoot tips on an osmotic solution). Cryopreservation is well advanced for vegetatively propagated species, and techniques are ready for large-scale experimentation in an increasing number of cases (Engelmann, 2011). We have started the routine cryopreservation with Czech potato genotypes in collaboration with Potato Research Institute Ltd., Havlíčkův Brod. The other three crops are supported by specialized companies (*Allium* genera by the Gene Bank Olomouc branch, the Research and Breeding Institute of Pomology at Holovousy and the Hop Research Institute). The Czech Plant Cryobank operates as a safe duplicate to repositories of germplasm kept in field or *in vitro* conditions. In this way the Czech Plant Cryobank in Prague joined the effort of the potato cryobank in Germany (Keller *et al.*, 2008), and Korea (Kim *et al.*, 2006). Currently three of the EU countries (The Czech Republic, Germany and Poland) involved in the maintenance of national vegetatively propagated *Allium* collections are developing the methodology for cryopreservation (project

EURALLIVEG, EU) and established the Tripartite *Allium* Cryobank to store meristematic explants in liquid nitrogen.

10. Acknowledgment

This work was partially supported by project 0002700604 MSMT and QH71228 MZe. The authors would like to thank M. Grospietsch, Jana Chittendenová, Aneta Babická and Adéla Hreňuková for their help and Dr. V. Skládal for critical review of this manuscript.

11. References

Benson, E.B.; Reed, B.M.; Brennan, R.M.; Clacher, K.A. & Ross, D.A. (1996). Use of thermal analysis in the evaluation of cryopreservation protocols for Ribes nigrum L germplasm. *Cryo-Letters*, Vol.17, No.6, (November 1996), pp. 347-362, ISSN0143-2044

Benson, E.E. (2008). Cryopreservation of phytodiversity: A critical appraisal of theory practice. *Critical Reviews in Plant Sciences*, Vol.27, No.3, (2008), pp. 141-219, ISSN0735-2689

Chang, Y.J. & Reed, B.M. (2001). Preculture conditions influence cold hardiness and regrowth of Pyrus cordata shoot tips after Cryopreservation. *Hortscience*, Vol.36, No.7, (December 2001), pp. 1329-1333, ISSN0018-5345

Condello, E.; Caboni, E.; Andre, E.; Piette, B.; Druart, P.; Swennen, R. & Panis, B. (2011). Cryopreservation of Apple in Vitro Axillary Buds Using Droplet-Vitrification. *Cryoletters*, Vol.32, No.2, (March 2011), pp. 175-185, ISSN0143-2044

Dereuddre, J.; Hassen, N.; Blandin, S. & Kaminski, M. (1991). Resistance of Alginate-Coated Somatic Embryos of Carrot (Daucus-Carota L) to Desiccation and Freezing in Liquid-Nitrogen .2. Thermal-Analysis. *Cryo-Letters*, Vol.12, No.3, (May 1991), pp. 135-148, ISSN0143-2044

Dumet, D.; Grapin, A.; Bailly, C. & Dorion, N. (2002). Revisiting crucial steps of an encapsulation/desiccation based cryopreservation process: importance of thawing method in the case of Pelargonium meristems. *Plant Science*, Vol.163, No.6, (December 2002), pp. 1121-1127, ISSN0168-9452

Ellis, D.; Skogerboe, D.; Andre, C.; Hellier, B. & Volk, G. (2006). Implementation of garlic cryopreservation techniques in the national plant germplasm system. *Cryoletters*, Vol.27, No.2, (March 2006), pp. 99-106, ISSN0143-2044

Engelmann, F. (2011). Use of biotechnologies for the conservation of plant biodiversity. *In Vitro Cellular & Developmental Biology-Plant*, Vol.47, No.1, (February 2011), pp. 5-16, ISSN1054-5476

Fabre, J. & Dereuddre, J. (1990). Encapsulation Dehydration - A New Approach to Cryopreservation of Solanum Shoot-Tips. *Cryo-Letters*, Vol.11, No.6, (November 1990), pp. 413-426, ISSN0143-2044

Fahy, G.M. (1985). Cold Shock Injury Is A Significant Factor in Freezing-Injury - A Position Against. *Cryobiology*, Vol.22, No.6, (1985), pp. 628-628, ISSN0011-2240

Faltus, M. & Zámečník, J. (2009). Thermal Characteristics of Some Vitrification Solutions. *Cryoletters*, Vol.30, No.5, (September 2009), pp. 389-389, ISSN0143-2044

Gonzalez-Arnao, M.T.; Panta, A.; Roca, W.M.; Escobar, R.H. & Engelmann, F. (2008). Development and large scale application of cryopreservation techniques for shoot and somatic embryo cultures of tropical crops. *Plant Cell Tissue and Organ Culture*, Vol.92, No.1, (January 2008), pp. 1-13, ISSN0167-6857

Grospietsch, M.; Stodůlková, E. & Zámečník, J. (1999). Effect of osmotic stress on the dehydration tolerance and cryopreservation of Solanum tuberosum shoot tips. *Cryo-Letters*, Vol.20, No.6, (November 1999), pp. 339-346, ISSN0143-2044

Gupta, S. & Reed, B.M. (2006). Cryopreservation of shoot tips of blackberry and raspberry by encapsulation-dehydration and vitrification. *Cryoletters*, Vol.27, No.1, (January 2006), pp. 29-42, ISSN0143-2044

Halmagyi, A.; Fischer-Kluver, G.; Mix-Wagner, G. & Schumacher, H.M. (2004). Cryopreservation of Chrysanthemum morifolium (Dendranthema grandiflora Ramat.) using different approaches. *Plant Cell Reports*, Vol.22, No.6, (January 2004), pp. 371-375, ISSN0721-7714

Hirai, D. & Sakai, A. (1999). Cryopreservation of in vitro-grown meristems of potato (Solanum tuberosum L.) by encapsulation-vitrification. *Potato Research*, Vol.42, No.2, (1999), pp. 153-160, ISSN0014-3065

Kaczmarczyk, A. (2008). Physiological, biochemical ,histological and ultrastructural aspects of cryopreservation in meristematic tissue of potato shoot tips (2008), ISSN 978-3-89574-694-9

Kaczmarczyk, A., Rokka, V-M & Keller E. R. J. (2011). Potato Shoot Tip Cryopreservation. A Review Source: *Potato research*, Vol. 54, No. 1, (March 2011),pp. 45-79, ISSN0014-3065

Kartha, K.K. & Leung, N.L. (1979). Cryopreservation of Plant Meristems As A Means of Germ Plasm Storage. *Cryobiology*, Vol.16, No.6, (1979), pp. 582-583, ISSN0011-2240

Keller, E.R.J.; Kaczmarczyk, A. & Senula, A. (2008). Cryopreservation for plant genebanks - A matter between high expectations and cautious reservation. *Cryoletters*, Vol.29, No.1, (January 2008), pp. 53-62, ISSN0143-2044

Kim, H.H.; Cho, E.G.; Baek, H.J.; Kim, C.Y.; Keller, E.R.J. & Engelmann, F. (January 2004). Cryopreservation of garlic shoot tips by vitrification: Effects of dehydration, rewarming, unloading and regrowth conditions. *Cryoletters*, Vol.25, No.1, (January 2004), pp. 59-70, ISSN0143-2044

Kim, H.H.; Lee, J.K.; Yoon, J.W.; Ji, J.J.; Nam, S.S.; Hwang, H.S.; Cho, E.G. & Engelmann, F. (2006). Cryopreservation of garlic bulbil primordia by the droplet-vitrification procedure. *Cryoletters*, Vol.27, No.3, (May 2006), pp. 143-153, ISSN0143-2044

Kim, H.H.; Lee, Y.G.; Shin, D.J.; Ko, H.C.; Gwag, J.G.; Cho, E.G. & Engelmann, F. (2009). Development of Alternative Plant Vitrification Solutions in Droplet-Vitrification Procedures. *Cryoletters*, Vol.30, No.5, (September 2009), pp. 320-334, ISSN0143-2044

Langis, R. & Steponkus, P.L. (March 1990). Cryopreservation of Rye Protoplasts by Vitrification. *Plant Physiology*, Vol.92, No.3, (March 1990), pp. 666-671, ISSN0032-0889

Liu, H.Q.; Yu, W.G.; Dai, J.X.; Gong, Q.H.; Yang, K.F. & Lu, X.Z. (2004a). Cryopreservation of protoplasts of the alga Porphyra yezoensis by vitrification. *Plant Science*, Vol.166, No.1, (January 2004a), pp. 97-102, ISSN0168-9452

Liu, Y.G.; Wang, X.Y. & Liu, L.X. (2004b). Analysis of genetic variation in surviving apple shoots following cryopreservation by vitrification. *Plant Science*, Vol.166, No.3, (2004b), pp. 677-685, ISSN0168-9452

Luyet, B. (1937). On the mechanism of cellular death by high pressure cytological modifications accompanying death in yeast. *Comptes Rendus Hebdomadaires Des Seances De l Academie Des Sciences*, Vol.204, (1937), pp. 1506-1508, ISSN0001-4036

Martinez, D. & Revilla, M.A. (1998). Cold acclimation and thermal transitions in the cryopreservation of hop shoot tips. *Cryo-Letters*, Vol.19, No.6, (November 1998), pp. 333-342, ISSN0143-2044

Martinez, D.; Revilla, M.A.; Espina, A.; Jaimez, E. & Garcia, J.R. (1998). Survival cryopreservation of hop shoot tips monitored by differential scanning calorimetry. *Thermochimica Acta*, Vol.317, No.1, (July 1998), pp. 91-94, ISSN0040-6031

Martinez, D.; Tames, R.S. & Revilla, M.A. (1999). Cryopreservation of in vitro-grown shoot-tips of hop (Humulus lupulus L.) using encapsulation/dehydration. *Plant Cell Reports*, Vol.19, No.1, (November 1999), pp. 59-63, ISSN0721-7714

Matsumoto, T. & Sakai, A. (1995). An Approach to Enhance Dehydration Tolerance of Alginate-Coated Dried Meristems Cooled to -196-Degrees-C. *Cryo-Letters*, Vol.16, No.5, (September 1995), pp. 299-306, ISSN0143-2044

Murashige, T. & Skoog, F. (1962). A Revised Medium for Rapid Growth and Bio Assays with Tobacco Tissue Cultures. *Physiologia Plantarum*, Vol.15, No.3, (1962), pp. 473-&, ISSN0031-9317

Niino, T. & Sakai, A. (1992). Cryopreservation of Alginate-Coated Invitro-Grown Shoot Tips of Apple, Pear and Mulberry. *Plant Science*, Vol.87, No.2, (1992), pp. 199-206, ISSN0168-9452

Niino, T.; Sakai, A.; Yakuwa, H. & Nojiri, K. (1992). Cryopreservation of In vitro-Grown Shoot Tips of Apple and Pear by Vitrification. *Plant Cell Tissue and Organ Culture*, Vol.28, No.3, (March 1992), pp. 261-266, ISSN0167-6857

Nishizawa, S.; Sakai, A.; Amano, Y. & Matsuzawa, T. (1993). Cryopreservation of Asparagus (*Asparagus-Officinalis* L) Embryogenic Suspension Cells and Subsequent Plant-Regeneration by Vitrification. *Plant Science*, Vol.91, No.1, (1993), pp. 67-73, ISSN0168-9452

Paulet, F.; Engelmann, F. & Glaszmann, J.C. (1993). Cryopreservation of Apices of In-Vitro Plantlets of Sugarcane (*Saccharum* Sp Hybrids) Using Encapsulation Dehydration. *Plant Cell Reports*, Vol.12, No.9, (July 1993), pp. 525-529, ISSN0721-7714

Reed, B.M.; Schumacher, L.; Wang, N.; D'Achino, J. & Barker, R.E. (2006). Cryopreservation of bermudagrass germplasm by encapsulation dehydration. *Crop Science*, Vol.46, No.1, (January 2006), pp. 6-11, ISSN0011-183X

Sakai, A. & Engelmann, F. (2007). Vitrification, encapsulation-vitrification and droplet-vitrification: A review. *Cryoletters*, Vol.28, No.3, (May 2007), pp. 151-172, ISSN0143-2044

Sakai, A.; Kobayashi, S. & Oiyama, I. (1991). Survival by Vitrification of Nucellar Cells of Navel Orange (*Citrus-Sinensis* Var Brasiliensis Tanaka) Cooled to 196-Degrees-C. *Journal of Plant Physiology*, Vol.137, No.4, (February 1991a), pp. 465-470, ISSN0176-1617

Sarkar, D. & Naik, P.S. (1998). Cryopreservation of shoot tips of tetraploid potato (*Solanum tuberosum* L.) clones by vitrification. *Annals of Botany*, Vol.82, No.4, (October 1998), pp. 455-461, ISSN0305-7364

Schafer-Menuhr, A.; Muller, E. & Mix-Wagner, G. (1996). Cryopreservation: an alternative for the long-term storage of old potato varieties. *Potato Research*, Vol.39, No.4, (1996), pp. 507-513, ISSN0014-3065

Sestak, J. & Zamecnik, J. (2007). Can clustering of liquid water and thermal analysis be of assistance for better understanding of biological germplasm exposed to ultra-low temperatures. *Journal of Thermal Analysis and Calorimetry*, Vol.88, No.2, (May 2007), pp. 411-416, ISSN1388-6150

Seufferheld, M.J.; Stushnoff, C.; Forsline, P.L. & Gonzalez, G.H.T. (1999). Cryopreservation of cold-tender apple germplasm. *Journal of the American Society for Horticultural Science*, Vol.124, No.6, (November 1999), pp. 612-618, ISSN0003-1062

Suzuki, M.; Tandon, P.; Ishikawa, M. & Toyomasu, T. (2008). Development of a new vitrification solution, VSL, and its application to the cryopreservation of gentian axillary buds. *Plant Biotechnology Reports*, Vol.2, No.2, (June 2008), pp. 123-131, ISSN1863-5466

Tanaka, D.; Niino, T.; Isuzugawa, K.; Hikage, T. & Uemura, M. (2004). Cryopreservation of shoot apices of in-vitro grown gentian plants: Comparison of vitrification and encapsulation-vitrification protocols. *Cryoletters*, Vol.25, No.3, (May 2004), pp. 167-176, ISSN0143-2044

Towill, L.E. (1990). Cryopreservation of Isolated Mint Shoot Tips by Vitrification. *Plant Cell Reports*, Vol.9, No.4, (August 1990), pp. 178-180, ISSN0721-7714

Tyler, N.; Stushnoff, C. & Gusta, L.V. (1988). Freezing of Water in Dormant Vegetative Apple Buds in Relation to Cryopreservation. *Plant Physiology*, Vol.87, No.1, (May 1988), pp. 201-205, ISSN0032-0889

Uragami, A.; Sakai, A.; Nagai, M. & Takahashi, T. (1989). Survival of Cultured-Cells and Somatic Embryos of Asparagus-Officinalis Cryopreserved by Vitrification. *Plant Cell Reports*, Vol.8, No.7, (October 1989), pp. 418-421, ISSN0721-7714

Wu, Y.J.; Engelmann, F.; Zhao, Y.H.; Zhou, M.D. & Chen, S.Y. (1999). Cryopreservation of apple shoot tips: Importance of cryopreservation technique and of conditioning of donor plants. *Cryo-Letters*, Vol.20, No.2, (March 1999), pp. 121-130, ISSN0143-2044

Wu, Y.J.; Zhao, Y.H.; Engelmann, F.; Zhou, M.D.; Zhang, D.M. & Chen, S.Y. (2001). Cryopreservation of apple dormant buds and shoot tips. *Cryoletters*, Vol.22, No.6, (November 2001), pp. 375-380, ISSN0143-2044

Zámečník, J. & Faltus, M. (2009). Evaluation of Thermograms from Differential Scanning Calorimeter. *Cryoletters*, Vol.30, No.5, (September 2009), pp. 388-388, ISSN0143-2044

Zámečník, J, Faltus, M. & Bilavčík, A. (2007). Cryoprotocols used for cryopreservation of vegetatively propagated plants in the Czech cryobank. *Advances in Horticultural Science*, Vol.21, (2007), pp. 247-250,

Zamecnik, J.; Faltus, M.; Kotkova, R. & Hejnak, V. (2011). Glass transition determination in Allium shoot tips after dehydration. Acta Hort. (ISHS) 908 (April 2009), pp. 33-38, ISSN0567-7572

Zhao, Y.H.; Wu, Y.J.; Engelmann, F. & Zhou, M.D. (2001). Cryopreservation of axillary buds of grape (*Vitis vinifera*) in vitro plantlets. *Cryoletters*, Vol.22, No.5, (September 2001), pp. 321-328, ISSN0143-2044

Zhao, Y.H.; Wu, Y.J.; Engelmann, F.; Zhou, M.D.; Zhang, D.M. & Chen, S.Y. (1999). Cryopreservation of apple shoot tips by encapsulation-dehydration: Effect of preculture, dehydration and freezing procedure on shoot regeneration. *Cryo-Letters*, Vol.20, No.2, (March 1999), pp. 103-108, ISSN0143-2044

10

Proline and the Cryopreservation of Plant Tissues: Functions and Practical Applications

David J. Burritt

The Department of Botany,
The University of Otago,Dunedin
New Zealand

1. Introduction

Cryopreservation has been proven to be an effective technology for the cost-effective, long-term preservation of genetic material. A wide range of plant material including cultured cells, tissues, embryos, meristems, pollen and seeds can be effectively preserved for extended periods of time and, when thawed, can be used to rapidly produce stock plants, with good preservation of genetic and physiological characteristics. Numerous protocols including controlled rate cooling, PVS2 vitrification, encapsulation-vitrification, and encapsulation-dehydration have been developed that allow the cryopreservation of a wide range of plant germplasm (Burritt, 2008), but irrespective of the protocol used each step in a cryopreservation protocol has the potential to impose a stress on plant cells. Low temperatures that lead to freezing can impose stress on cells and tissues in two ways, by the direct effects of low temperatures on cell function and integrity or by the cellular dehydration that occurs when the cells water freezes. Several of the mechanisms by which these two forms of stress can damage plant cells are shown in Figure 1.

Numerous studies have shown that cold temperatures induce the accumulation of metabolites, including low-molecular-weight carbohydrates such as fructose, glucose, maltose and raffinose, and amino acids such as proline and glutamine (Taji et al., 2002; Cook et al., 2004). These metabolites play important protective roles in freezing tolerance in whole plants (Kaplan and Guy, 2004) and this has lead to their extensive use in the protocols developed for the cryopreservation of isolated plant cells and tissues (Burritt, 2008). In particular, the amino acid proline has been found to help confer freezing tolerance in a wide variety of both animal and plant cells, and is often added to cryoprotective solutions or is used for preconditioning plants or pretreating isolated cells or tissues prior to cryopreservation (Burritt, 2008). Despite its widespread use, little is known of the mechanisms via which proline protects cells during cryopreservation.

This chapter gives an overview of proline synthesis and metabolic regulation in plants and the changes in proline metabolism associated with desiccation and freezing tolerance, which are both of importance for the successful cryopreservation of plants cells and tissues. The use of proline as a cryoprotectant or pre-growth additive for the cryopreservation of plant cells and tissues is then overviewed and the potential mechanisms via which proline can protect plant cells is critically evaluated. Future research needs are then discussed.

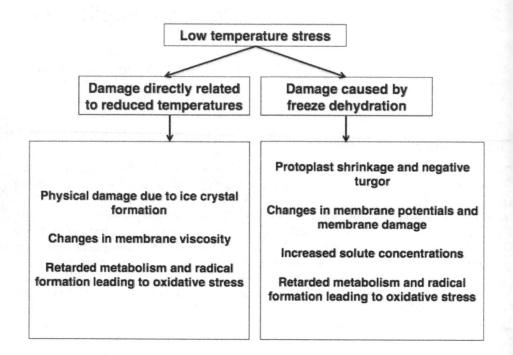

Fig. 1. Potential damage caused by the stresses associated with exposure of plant cells to low temperatures.

2. Proline and plants

2.1 The function of proline in plants

Essential for primary metabolism, both as a free amino acid and as a component of proteins, proline is distinctive among the proteinogenic amino acids as it contains a secondary amino group and a distinctive cyclic structure (Lehmann et al., 2010). The cyclic structure of proline causes exceptional conformational rigidity, compared to other amino acids, as proline's side chain locks its φ backbone dihedral angle at approximately -75° and this determines the arrangement of the peptide chain and can lead to the stabilization or destabilization of secondary protein structures.

As well as being important for primary metabolism proline appears to have numerous other functions in plants. Research has clearly demonstrated that proline levels show significant fluctuations in response to environmental stress (Bohnert et al., 1995), but the precise mode of action of proline remains largely a matter of speculation. In addition to its role in environmental stress tolerance, recent research has provided evidence that proline may also play important roles in plant development both as a metabolite and as a signal molecule (Mattioli et al., 2009). Studies have shown that proline could play important roles in embryo and seed development, stem elongation, and the transition from vegetative growth to flowering (Mattioli et al., 2008; Mattioli et al., 2009)

2.3 Proline biosynthesis and catabolism

The proposed pathways for proline biosynthesis and catabolism in plants are outlined in Figure 2. In plants proline can be synthesized from glutamate or ornithine, however under most conditions proline is mainly synthesized from glutamate rather than from ornithine, as the enzyme ornithine-δ –aminotransferase (dOAT) is down regulated (Szabados & Savoure, 2009). Two enzymes are required for the synthesis of proline from glutamate. The first enzyme, $\Delta 1$- pyrroline-5-carboxylate synthase (P5CS) is a bifunctional enzyme that phosphorylates and reduces glutamate to glutamyl-5- semialdehyde (G5SA) that then spontaneously converts to $\Delta 1$- pyrroline-carboxylate (P5C). The second enzyme, $\Delta 1$-pyrroline-carboxylate reductase (P5CR) further reduces the P5C intermediate to proline (Delauney & Verma, 1993). P5CS has been found to be encoded by 2 genes in most plants, while P5C is encoded by only a single gene (Szekely et al. 2008; Strizhov et al., 1997) The rate-limiting step in the above pathway is the γ -glutamyl kinase activity of P5CS, which is sensitive to feedback inhibition by the presence of relatively low cellular proline levels (Zhang et al., 1995). Alternatively proline can be synthesized from ornithine by dOAT, which converts ornithine and α-ketoglutarate to P5C and glutamate by transamination (Stranska et al., 2008). Funck et al., 2008, in a study of *Arabidopsis thaliana,* found that mutant plants which lacked dOAT activity could not mobilize nitrogen from arginine or ornithine, but could accumulate proline and so suggested the main role for dOAT was arginine degradation. They also suggested that as dOAT is localized in the mitochondria and that it would be unlikely that P5CR could directly utilize dOAT-generated P5C, as P5CR is localized in the cytosol or in plastids.

Proline degradation in plants takes place in mitochondria and so is by in large separated from the biosynthetic pathway. The first step in proline catabolism is the oxidation of proline to P5C by proline dehydrogenase (PDH), which in Arabidopsis and tobacco is encoded by two homologous genes (Mani et al., 2002; Ribarits et al., 2007; Verbruggen & Hermans 2008). The P5C generated is then converted to glutamate by pyrroline-5-carboxylate dehydrogenase (P5CDH), which is thought to be encoded by a single gene in all of the plant species analysed to date (Ayliffe et al. 2005; Mitchell et al. 2006). However, biochemical analysis P5CDH in *Nicotiana plumbaginifolia* and *Zea mays* has revealed two slightly different enzyme activities that may arise from a single gene, or a second P5CDH gene may be present (Elthon & Stewart 1982; Forlani et al. 1997). In plants under stress, the accumulation of proline is thought to be due not only to increased synthesis, but also to inactivation of degradation pathways (Delauney & Verma, 1993)).

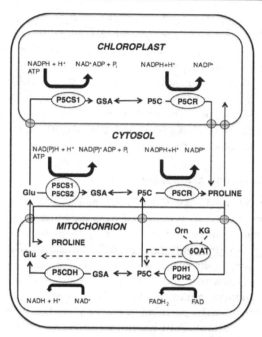

Fig. 2. Proposed model for proline metabolism in higher plants (adapted from Lehmann et al. 2010). Glu glutamate, Orn ornithine, P5C pyrroline-5-carboxylate, GSA glutamic-γ-semialdehyde, KG α-ketoglutarate. P5CS P5C synthetase, P5CR P5C reductase, PDH proline dehydrogenase, P5CDH P5C dehydrogenase, δOAT ornithine-δdaminotransferase. Transporters and potential transporters are shown as grey circles.

3. Proline in plant cells under stress

3.1 Proline accumulation in higher plants

Stress has been shown to induce proline accumulation in a wide range of organisms including eubacteria, protozoa, invertebrates and plants (Verbruggen & Hermans, 2008; Kostal et al., 2011) and proline accumulation is believed to be very important as part of the physiological adaptation of plants to stress. In plants a wide range of abiotic and biotic stressors have been shown to induce proline accumulation including, salt, drought, high temperatures, low temperatures, heavy metals, anaerobiosis, nutrient deficiency, organic pollutants, ultraviolet (UV) radiation and pathogen infection (Chu et al., 1978; Alia & Saradhi, 1991; Saradhi et al., 1995; Hare et al., 1999; Siripornadulsil et al., 2002). The level of proline that accumulates in plants in response to stress varies greatly and is highly dependent on the plant species, with increase of up to 100 fold compared to controls reported in the literature (Verbruggen & Hermans, 2008).

With respect to cryopreservation, numerous studies have demonstrated the importance of proline for plant cold tolerance (Swaaij, Jacobsen & Feenstra 1985, Swaaij et al. 1986; Duncan & Widholm 1987; Ait-Barka & Audran 1997; Hoffman et al., 2010; Javadian et al., 2010; Burbulis et al., 2011). Studies on plants relatively insensitive to chilling, such as barley (Chu et al. 1978), rye (Koster & Lynch 1992), winter wheat (Dorffling et al. 1997), and *Arabidopsis thaliana* (Xin & Browse 1998; Nanjo et al. 1999) have demonstrated significant positive correlations between cellular proline accumulation and improved cold tolerance.

In addition, plant cells under dehydrating conditions, which are often a consequnce of cryopreservation, undergo osmotic adjustment by accumulating one or several low molecular weight organic solutes, which are often referred to as compatible osmolytes and/or osmoprotectants. These molecules play a critical role in counteracting the effect of osmotic stress in plants at the cellular level (Yoshiba et al., 1997). In plants under dehydrating conditions such as drought or high salinity, proline is one of the most common compatible osmolytes and while several amino acids are known to accumulate in response to osmotic stress, proline appears to be the preferred organic osmoticum in many plants and may have a specific protective role in the adaptation of plant cells to dehydration. For example, in a study of *Triticum aestivum* L. (durum wheat) under salinity stress, Poustini et al. (2007) found a positive correlation between proline levels and osmotic potential, and concluded that proline is an important osmolyte for osmotic adjustment in wheat under water stress. In addition, it has been demonstrated that transgenic tobacco plants with elevated levels of proline biosynthesis show increased tolerance to hyperosmotic stress (Kavi Kishot et. al., 1995), providing further evidence of a cause-and-effect relationship between proline levels and osmotic tolerance. Proline normally accumulates in the cytosol, where it contributes to the cytoplasmic osmotic adjustment in response to water loss without interfering with normal cellular processes and biochemical reactions (Ashraf & Foolad, 2007).

3.2 Proline and cryopreservation

During cryopreservation, plant cells encounter similar problems to those they encounter under freezing conditions in the field. They under go changes in the spatial organization of biological membranes, biochemical and chemical reactions can be retarded, and the status and availability of water can be altered. For these reasons proline is likely to be an effective cryoprotectant for cryopreserved plant cells and tissues.

4. The use of proline as a cryoprotectant

Proline has been used for many years in numerous cryoprotection protocals for the preservation of a wide range of both animal and plant cells and tissues. For example, Li et. al. (2003) investigated the effects of addition of proline, glutamine, and glycine to the Tes-Tris-egg yolk (TTE) freezing medium used for cryopreservation of cynomolgus monkey (*Macaca fascicularis*) spermatozoa. They found that the addition of 5 mM proline, 10 mM glutamine, and 10 or 20 mM glycine to TTE significantly improved post-thaw sperm motility and membrane integrity compared to controls without an amino acid. Of the three amino acids tested proline was effective at the lowest concentration.

Proline has also been found to be useful for the cryopreservation of plant cells, meristems and embryos. Jain et al. (1996) included proline in the cryoprotectant solution as part of a protocol that was used to successfully cryopreserve embryogenic suspension cells of two commercially cultivated aromatic Indica rice varieties using a simple one-step freezing procedure that did not require a controlled-rate freezer. Brison et al. (1995) used a preculture medium enriched with dimethylsulfoxide and proline prior to the cropreservation of *in vitro* grown interspecific Prunus rootstock, Fereley-Jaspi (R). In a study to develop a cryoprotection protocol for highly freezing sensitive *Begonia* species, Burritt (2008) found that adventitious shoots of the rhizomatous begonia, *Begonia x erythrophylla* were sensitive to dehydration and very sensitive to freezing. While pre-treatment with 0.75 M sucrose significantly increased the percentage of encapsulated shoots surviving dehydration, pre-treatment with sucrose did not afford cryoprotection without prior dehydration.

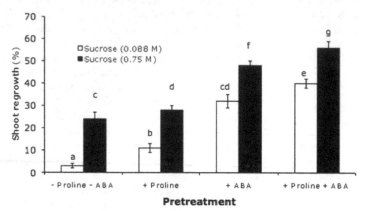

Fig. 3. The percentage of *Begonia x erythrophylla* shoots surviving pre-treatment with ABA (3.8 uM) and/or proline (2.15 mM). The different letters indicate statistically different values at p < 0.05 (modified from Burritt 2008).

Addition of abscisic acid (ABA) and proline to the pre-treatment medium significantly improved the percentage of shoots surviving freezing. Pre-treatment of shoots with a medium containing, 0.75M sucrose, 3.8 µM ABA and 2.15 mM proline resulted in greater than 50% of shoots surviving freezing (Figure 3).

Christianson (1998) used a 3-4 day preconditioning treatment using a tissue culture medium supplemented with 10^{-5} M ABA and 100 mM proline to greatly increase survival rates and simplify a protocol for moss cryopreservation. Pretreatment with the combination of proline and ABA was used as part of a cryopreservation protocol that could be used for *Ceratodon purpureus*, *Funaria hygrometrica*, *Physcomitrella patens*, and two species of *Sphagnum*. Cryopreserved cultures remained viable at least one year at -80ºC.

In addition to both animal and plant cells, proline has been found to be particularily useful for the cryopreservation of algal cells. Kuwano et al. (2004) found the gametophytic cells of six species of Laminariales, *Laminaria japonica* Areschoug, *L. longissima* Miyabe, *Kjellmaniella crassifolia* Miyabe, *Ecklonia stolonifera* Okamura, *E. kurome* Okamura, and *Undaria pinnatifida* (Harvey) Suringar could be cropreserved using a cryoprotective solution containing ethylene glycol and proline. The cells were suspended in a mixture of ethylene glycol and proline, and slowly cooled to -40ºC over a period of 4 h. After a cooling step, the cells were immediately immersed in liquid nitrogen. Viabilities ranged from 36.2% to 67.2%. Nanb et al. (2009) developed a cryopreservation protocol for gametophyte strains of the edible macroalgae *Undaria pinnatifida* (Harvey). Following a pretreatment involving exposure of male and female gametophytes to low levels of light, they used a two-step cooling method with a mixture of cryoprotectants including 10% L-proline and 10% glycerol, before freezing in liquid nitrogen. Gametophyte survival rates were high, ranging from 43-60% for females and 64-100% for males. The morphology of the sporophytes formed from the cryopreserved gametophytes appeared normal and the authors suggested that this cryopreservation method could be used to preserve culture stocks of U. *pinnatifida* for mariculture.

5. Proline a multifunctional cryoprotectant?

5.1 Possible mechanisms of protection

Because of its ability to act as an osmoprotectant without interfering with normal cellular processes and biochemical reactions proline has been used in a range of different cryopreservation protocols both for animal and plants cells and tissues, however the exact mode by which protection is achieved is still a matter of considerable debate in the scientific literature. Proline could potentially acting as storage reserve of carbon and nitrogen, a compatible osmolyte, a buffer for cytosolic pH, a scavenger of reactive oxygen species (ROS) and as an aid to balancing cellular redox status (Smirnoff & Cumbes 1989; Hare & Cress, 1997). It has also been proposed that proline could act as a molecular chaperone, helping to stabilize the structure of proteins, and as part of the signal transduction chain alerting plant cells to the presence of a stressor and hence triggering adaptive responses (Maggio et al. 2002).

5.2 Proline as an osmolite

The osmoregulatory role of proline in plant cells exposed to hyperosmotic stress has been the subject of numerous studies and under environmental conditions that result in cellular dehydration such as drought, freezing or extreme salinity, it is widely accepted that proline accumulates and acts as a compatible solute helping to protect cells from damage (Heur, 1994). Accumulation of cytoplasmic osmolytes, such as proline, is thought to aid in reducing the cellular water potential to a level below the external water potential, this enables water to move into the cell and be maintained there, while at the same time minimising potentially deleteriously increases in ionic strength. However, there is some debate in the published literature as to whether increased cytosolic levels of free proline has any direct adaptive value (Heur, 1994). While there are many reports of positive correlations between the capacity for proline accumulation and dehydration and cold tolerance (see section 3.1), some researchers still challenge the value of the ability of plant cells to accumulate proline as a positive index for osmotic stress resistance (Heur, 1994 & references therein).

5.3 Proline as precursor for other molecules

It has been suggested that stress-induced accumulation of amino acids like proline may not only have an osmoregulatory role, but that they could also be a mechanism to provide cells with a pool of the precursors required to synthesis other molecules known to be involved in biotic and abiotic stress responses (Sanchez et al., 2008). For example polyamines can be synthesized from arginine or ornithine and ornithine from glutamate, hence the pathways for proline and polyamine biosynthesis are interlinked, and both groups of molecules are important in plant stress responses (Groppa & Benavides 2008). Little is known about the roles of polyamine metabolism in the process of cryopreservation, but Ramon et al. (2002) reported that an increase in putrescine content was positively correlated with the survival rate after simple freezing or after vitrification of banana meristem cultures.

Stored amino acids could also be useful during the recovery process following stress. The accumulation of large cellular pools of amino acids could allow the rapid synthesis of enzymes and the repair of structural proteins, allowing a more rapid recovery of cells following cryopreservation, but this possibility has yet to investigated.

5.4 Proline as an antioxidant

Reactive oxygen species, such as the superoxide anion ($O_2 \bullet$), hydrogen peroxide (H_2O_2), and the extremely reactive hydroxyl radical ($\bullet OH$) are produced within cells as a consequence of normal metabolic processes, but the production of ROS often increases when cells are under stress (Smirnoff, 1993; Halliwell & Gutteridge, 1999). When ROS are produced at levels high enough to overcome the antioxidant defences that normally control cellular ROS levels, oxidation of DNA, proteins and membrane fatty acids occurs, the latter can result in lipid peroxidation and loss of membrane function (Halliwell & Gutteridge, 1999). Such damage is commonly referred to as oxidative stress (Lesser, 2006; Burritt & MacKenzie 2003; Burritt, 2008). Cryopreservation protocols comprise a number of steps, each of which has the potential to cause stress that could increase ROS production. Recent studies have shown that dehydration and freezing can both lead to increased ROS production and lead to oxidative stress (Feck et al., 2000; Roach et al., 2008). A recent study on oxidative stress and antioxidant metabolism during the cryopreservation of olive somatic embryos demonstrated the importance of oxidative stress and antioxidant metabolism for the successful cryopreservation of plant cells (Lynch et al., 2011).

As mentioned in Section 4 Burritt (2008) found that addition of ABA and proline to the pre-treatment medium significantly improved the percentage of *Begonia x erythrophylla* shoots surviving freezing, this increase in percentage survival was accompanied by a decrease in levels of hydrogen peroxide (Figure 4) and oxidative damage, measured as the levels of lipid peroxides, observed in the shoots immediately following thawing (Figure 5).

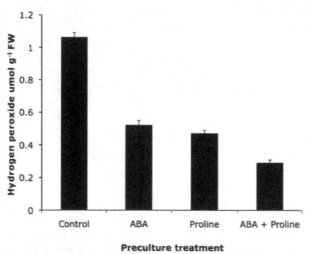

Fig. 4. The influence of pre-treatment with ABA (3.8 uM) and/or proline (2.15 mM) on hydrogen peroxide levels, as determined by Chesseman (2006), in post thaw *Begonia x erythrophylla* shoots cryopreserved as described by Burritt (2008).

Because of its chemical properties proline has a high capacity to quench singlet oxygen and hydroxyl radicals. Pyrrolidine, which forms the 5- membered ring of proline, has a low IP and so proline is able to form a charge-transfer complex, enabling it to quench singlet oxygen effectively. Proline can also react with hydroxyl radicals under hydrogen abstraction forming a stable radical (Matylsik, 2002). Therefore the accumulation of proline to high levels in plant cells under stress or plants cells treated with exogenous proline as part of a cryopreservation protocol could greatly increase the ROS scavenging capacity of said cells and reduce the potential for oxidative damage. In particular, as proline has the potential to reduce ROS levels it could help reduce oxidative damage to vital cellular macromolecules and hence stabilize proteins (Anjum, 2000),) DNA (Iakobashvil, 1999) and lipid membranes (Alia, 1991). The accumulation of proline-rich proteins and particularly proline residues in cellular proteins is thought to provide additional protection against oxidative stress (Matylsik, 2002). The increase in ROS scavenging capacity brought about by increased intracellular proline levels could be a key mechanism by which proline helps reduce the freezing and dehydration associated cellular damage associated with most cryopreservation protocols.

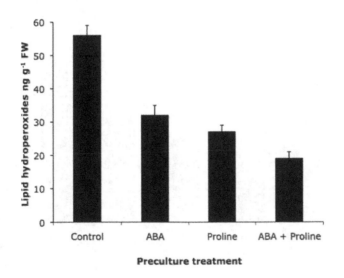

Fig. 5. The influence of pre-treatment with ABA (3.8 uM) and/or proline (2.15 mM) on lipid hydroperoxide levels, as determined by Mihaljevic et al. (1996), in post thaw *Begonia x erythrophylla* shoots cryopreserved as described by Burritt (2008).

Interestingly ABA combined with proline reduced hydrogen peroxide production and oxidative damage, measured as lipid peroxidation, more effectively in post thaw *Begonia x erythrophylla* shoots than ABA or proline alone. Christianson (1998) also found that ABA and proline in combination improved the survival of moss gametophytes following cryopreservation. These results suggest a possible interaction between ABA and proline may exist.

5.5 Is there an inaction between ABA and proline?

Studies have shown a relationship between proline and ABA with respect to cold tolerance (Xi & Li, 1993; Lou & Reid, 1997). In a recent study of maize suspension-cultured cells Chen and Li (2002) showed that an ABA treatment at warm temperatures improved the tolerance of cells to subsequent chilling, and that both ABA-treated and untreated maize cells accumulated proline in response to chilling. Chen and Li also found that ABA-treated cells showed less lipid peroxidation during chilling and unlike untreated cells were able to retain the accumulated proline intracellularly.

In post thaw *Begonia x erythrophylla* shoots ABA combined with proline resulted in much higher shoot survival than pretreatment with ABA or proline alone. Interestingly ABA combined with proline resulted in far higher intracellular proline concentrations (Figure 6). The greater concentrations of proline seen in the combined treatment could be due increased endogenous synthesis of proline, induced by exposure to ABA, combined with uptake of exogenous proline during the pretreatment phase and/or to an ABA induced mechanism that helps reduce proline leakage, but further investigations are required to determine how the combined application of ABA and proline increase shoot survival after cryopreservation.

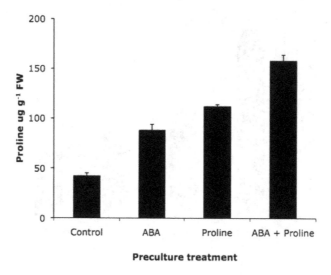

Fig. 6. The influence of pre-treatment with ABA (3.8 uM) and/or proline (2.15 mM) on proline levels, in post thaw *Begonia x erythrophylla* shoots cryopreserved as described by Burritt (2008).

5.5 Proline and direct macromolecule protection

As well as the potential protective mechanisms detailed above, proline has been shown to directly protect key cellular macromolecules, in particular lipid membranes and proteins such as enzymes (Verbruggen & Hermans, 2008). Proline molecules can intercalate between the head groups of membrane phospholipids during freeze-dehydration helping to reduce mechanical stresses in the membranes, or alter the physical properties of membranes making them less prone to a liquid crystalline-to-gel transition (Hoekstra et al., 2001). It has also been suggested that proline molecules can directly replace missing water molecules between the phospholipids headgroups (Rudolph et al., 1986).

In addition, according to the preferential exclusion hypothesis, proline is one of a group of solutes that, when in aqueous solution, are excluded from contact with the surfaces of proteins and phospholipid bilayers (Arakawa & Timasheff, 1983). Accordingly addition of proline to a solution stabilizes the native structure of protein monomers and protects oligomeric protein complexes from denaturation and dissociation. Rudolph et al. (1986) demonstrated that the activity of the enzyme lactate dehydrogenase could be protected in part during freeze-thaw cycles by increasing the concentration of proline from 0 to 200 mM in the buffer in which the enzyme was solubilised.

5.6 Other mechanisms

There are several other mechanisms via which proline could contribute to over coming the stresses associated with cryopreservation. For example, the accumulation of proline could also be a mechanism to store energy as the oxidation of a single proline molecule can produce up to 30 ATP equivalents (Atkinson, 1971). Replenishment of NADP+ and redox cycling have also been sugested as potential mechanisms associated with stress tolerance, (Hare & Cress 1999), as has a role in stress signal transdution (Hare et al., 1997).

6. Conclusions

While numerous studies have demonstrated that proline can be used to improve the survival of plant cells and organs following cryopreservation, there is little definitive evidence as to the mode of action of proline. More research is required to determine how proline protects plant cells at the cellular level and to determine how other treatments that confer cryotolerance, such as ABA pretreatments interact with proline metabolism and could hence improve the cryotolerance of plant cells. However, despite our lack of knowledge with respect to the mode of action of proline, this amino acid continues to be of great value as a cryoprotectant that can be used with a wide range of cell types from many different organisms.

6. References

Ait-Barka, E.; Audran, J.C. (1997) Response of champenoise grapevine to low temperatures: changes of shoot and bud proline concentrations in response to low temperatures and correlations with freezing tolerance. *Journal of Horticultural Science*, Vol.72,577–582.

Alia; Saradhi, P.P. (1991) Proline accumulation under heavy-metal stress. *Journal of Plant Physiology*, Vol.138,554-558.

Anjum, F.; Rishi, V.; Ahmad, F. (2000) Compatibility of osmolytes with Gibbs energy of stabilization of proteins. *Biochimica et biophysica acta-protein structure and molecular enzymology*, Vol.1476,75-84.

Arakawa, T.; Timasheff, S.N. (1985) The stabilisation of proteins by osmolytes. *Biophysical Journal*, Vol.47,411–414.

Ashraf M.; Foolad M. R. (2007) Roles of glycine betaine and proline in improving plant abiotic stress resistance. *Environmental and Experimental Botany*, Vol.59,206-216.

Atkinson, D,E, (1977) Cellular Energy Metabolism and Its Regulation. New York: Academic Press.

Ayliffe, M.A.; Mitchell, H.J.; Deuschle, K.; Pryor, A.J.; (2005) Comparative analysis in cereals of a key proline catabolism gene. *Molecular Genetics & Genomics*, Vol.274,494–505.

Bohnert, H.J.; Nelson, D.E.; Jensen, R.G. (1995) Adaptations to environmental stresses. *The Plant Cell*, Vol.7,1099–1111.

Brison, M.; Deboucaud, M.T.; Dosba, F. (1995) Cryopreservation of in-vitro grown shoot tips of 2 interspecific prunus rootstocks. *Plant Science*, Vol.105,235-242.

Burbulis, N.; Jonytiene, V.; Kupriene,R.; Blinstrubiene, A. (2011) Changes in proline and soluble sugars content during cold acclimation of winter rapeseed shoots in vitro. *Journal of Food Agriculture & Environment*, Vol.9,371-374.

Burritt, D.J. (2008) Efficient cryopreservation of adventitious shoots of *Begonia x erythrophylla* using encapsulation-dehydration requires pretreatment with both ABA and proline. *Plant Cell Tissue and Organ Culture*, Vol.95,209-215.

Burritt, D.J.; MacKenzie, S. (2003). Antioxidant metabolism during the acclimation of *Begonia x erythrophylla* to high-light. *Annals of Botany*, Vol.91, 783-794.

Cheeseman, J.M. (2006) Hydrogen peroxide concentrations in leaves under natural conditions. *Journal of Experimental Botany*, Vol.57,2435-2444.

Chen, W.P.; Li, P.H. (2002) Membrane stabilization by abscisic acid under cold aids proline in alleviating chilling injury in maize (*Zea mays* L.) cultured cells. *Plant Cell and Environment*, Vol.25,955-962.

Christianson, M.L. (1998) A simple protocol for cryopreservation of mosses. *The Bryologist*, Vol.101,32-35.

Chu, T.M.; Jusaitis, M.; Aspinall, D.; Paleg, L.G. (1978) Accumulation of free proline at low temperatures. *Physiologia Plantarum*, Vol.43, 254–260.

Cook, D.; Fowler, S.; Fiehn, O.; (2004) A prominent role for the CBF cold response pathway in configuring the low-temperature metabolome of Arabidopsis. *Proceedings of the National Academy of Sciences of the United States of America*, Vol.101,15243-15248. *Crop Science*, Vol.50,1037-1047.

Delauney, A.J.; Verma, D.P.S. (1993) Proline biosynthesis and osmoregulation in plants. *Plant Journal*, Vol4,215–223.

Dorffling, K.; Dorffling, H.; Lesselich, G.; Luck, E.; Zimmermann, C.; Melz, G.; Jurgens, H.U. (1997) Heritable improvement of frost tolerance in winter wheat by in vitro selection of hydroxyproline-resistant proline overproducing mutants. *Euphytica*, Vol.93,1-10.

Duncan, D.R.; Widholm, J.M. (1987) Proline accumulation and its implication in cold tolerance of regenerable maize callus. *Plant Physiology*, Vol.83,703-708.

Elthon, T.E.; Stewart, C.R.; (1982) Proline oxidation in corn mitochondria: involvement of NAD, relationship to ornithine metabolism, and sidedness on the inner membrane. *Plant Physiology*, Vol.70,567–572.

Fleck, R.A.; Benson, E.E.; Bremner, D.H.; Day, J.G. (2000) Studies of free radical-mediated cryoinjury in the unicellular green alga *Euglena gracilis* using a non-destructive hydroxyl radical assay: A novel approach for developing protistan cryopreservation strategies. *Free Radical Research*, Vol.32,157-170.

Forlani, G.; Scainelli, D.; Nielsen, E. (1997) Two D1-pyrroline-5- carboxylate dehydrogenase isoforms are expressed in cultured *Nicotiana plumbaginifolia* cells and are differentially modulated during the culture growth cycle. *Planta*, Vol. 202,242.

Funck, D.; Stadelhofer, B.; Koch, W. (2008) Ornithine-d-aminotransferase is essential for arginine catabolism but not for proline biosynthesis. *BMC Plant Biology*, Vol.8,40.

Groppa M. D.; Benavides M. P. (2008). Polyamines and abiotic stress: Recent advances. Amino Acids Vol.34, 35-45.

Halliwell, B.; Gutteridge, J.M.C. (1999). Free Radicals in Medicine and Biology, Edition 3. Oxford: Oxford University Press.

Hare, P.D.; & Cress, W.A. (1997) Metabolic implications of stress-induced proline accumulation in plants. *Plant Growth Regulation*, Vol.21,79–102.

Hare, P.D.; Cress, W.; van Staden, J. (1999) Proline synthesis and degradation: a model system for elucidating stress-related signal transduction. *Journal of Experimental Botany*, Vol.50,413–434.

Heuer, B. (1994) Osmoregulatory role of proline in water and salt-stressed plants. In: Pessarakli, M. (ed) Handbook of Plant and Crop Stress, pp 363–381. New York: Marcel Dekker, Inc.

Hoekstra, F.A.; Golovina, E.A.; Buitink, J. (2001) Mechanisms of plant desiccation tolerance. *Trends in Plant Science*, Vol.6,431-438.

Hoffman, L.; DaCosta, M.; Ebdon, S.J.; Watkins, E. (2010) Physiological Changes during Cold Acclimation of Perennial Ryegrass Accessions Differing in Freeze Tolerance.

Iakobashvil, R.; Lapidot, A. (1999) Low temperature cycled PCR protocol for Klenow fragment of DNA polymerase I in the presence of proline. Nucleic Acids Research Vol.27,1566-1568.

Jain, S.; Jain, R.K.; Wu, R. (1996) A simple and efficient procedure for cryopreservation of embryogenic cells of aromatic Indica rice varieties. *Plant Cell Reports*, Vol.15,712-717.

Javadian, N.; Karimzadeh, G.; Mahfoozi, S.; Ghanati, F. (2010) Cold-induced changes of enzymes, proline, carbohydrates, and chlorophyll in wheat. *Russian Journal of Plant Physiology*, Vol.57,540-547.

Kaplan F.; Guy C.L. (2004) Beta-amylase induction and the protective role of maltose during temperature shock. *Plant Physiology*, Vol.135,1674-1684.

Kavi Kishor, P.B.; Hong, Z.; Miao, G.-H.; Hu, C.-A.A.; Verma D.P.S. (1995) Overexpression of Δ1-pyrroline-5-carboxylic acid synthetase increases proline production and confers osmotolerance in transgenic plants. *Plant Physiology*, Vol.108,1387–1394.

Kostal, V.; Zahradnickova, H.; Simek, P. (2011) Hyperprolinemic larvae of the drosophilid fly, *Chymomyza costata*, survive cryopreservation in liquid nitrogen. *Proceedings of the National Academy of Sciences of the United States of America*, Vol.108,13041-13046.

Koster, K.L.; Lynch, D.V. (1992) Solute accumulation and compartmentation during the cold acclimation of Puma rye. *Plant Physiology*, Vol.98,108–113.

Kuwano, K.; Kono, S.; Jo, Y.H.; Shin, J.A.; Saga, N. (2004) Cryopreservation of the gametophytic cells of Laminariales (Phaeophyta) in liquid nitrogen. *Journal of Phycology*, Vol.40,606-610.

Lehmann Silke; Funck Dietmar; Szabados Laszlo; (2010) Proline metabolism and transport in plant development. *Amino Acids*, Vol.39,949-962.

Lesser, M.P (2006). Oxidative stress in marine environments: biochemistry and physiological ecology. *Annual Review of Physiology*, Vol.68, 253-278.

Li, Y.H.; Si, W.; Zhang, X.Z.; Dinnyes, A.; Ji, W.Z. (2003) Effect of amino acids on cryopreservation of cynomolgus monkey (*Macaca fascicularis*) sperm. *American Journal of Primatology*, Vol.59,159-165.

Luo, J. and B.M. Reed. 1997. Abscisic-acid responsive protein, bovine serum albumin,and proline pretreatments improve recovery of in vitro currant shoot-tip meristems and callus preserved by vitrification. *Cryobiology*, Vol.34,240-250.

Lynch, P.T.; Siddika, A.; Johnston, J.W.; Trigwell, S.M.; Mehra, A.; Benelli, C.; Lambardi, M.; Benson, E.E. (2011) Effects of osmotic pretreatments on oxidative stress, antioxidant profiles and cryopreservation of olive somatic embryos. *Plant Science*, Vol.181,47-56.

Maggio, A.; Miyazaki, S.; Veronese, P.; Does proline accumulation play an active role in stress-induced growth reduction? *Plant Journal*, Vol.31,699-712.

Mani, S.; Van de Cotte, B.; Van Montagu, M.; Verbruggen, N. (2002) Altered levels of proline dehydrogenase cause hypersensitivity to proline and its analogs in Arabidopsis. *Plant Physiology*, Vol.128,73–83.

Mattioli, R.; Costantino, P.; Trovato, M.; (2009) Proline accumulation in plants: not only stress. *Plant Signaling & Behavior* Vol.4,1016–1018

Mattioli, R.; Marchese. D.; D'Angeli, S.; Altamura, M.; Costantino, P.; Trovato, M.; (2008) Modulation of intracellular proline levels affects flowering time and inflorescence architecture in *Arabidopsis*. *Plant Molecular Biology*, Vol.66,277–288 .

Matysik J; Alia; Bhalu B; Mohanty, P. (2002) Molecular mechanisms of quenching of reactive oxygen species by proline under stress in plants. *Current Science*, Vol.82,525-532.

Mihaljevic, B.; Katusin-Razem, B.; Razem, D. (1996) The reevaluation of the ferric thiocyanate assay for lipid hydroperoxides with special considerations of the mechanistic aspects of the response. *Free Radical Biology and Medicine*, Vol.21,53-63.

Mitchell, H.J.; Ayliffe, M.A.; Rashid, K.Y.; Pryor, A.J. (2006) A rust- inducible gene from flax (*fis1*) is involved in proline catabolism. *Planta* Vol.223,213–222.

Nanb, N.; Fujiwara, T.; Kuwano, K.; Ishikawa, Y.; Ogawa, H.; Kado, R. (2009) Effect of pre-incubation irradiance on survival of cryopreserved gametophytes of *Undaria pinnatifida* (Phaeophyta) and morphology of sporophytes formed from the gametophytes. *Aquatic Botany*, Vol.90,101-104.

Nanjo, T.; Kobayashi, M.; Yoshiba, Y.; Kakubari, Y.; Yamaguchi-Shi-nozaki, K.; Shinozaki, K. (1999) Antisense suppression of proline degradation improves tolerance to freezing and salinity in *Arabidopsis thaliana*. *FEBS Letters*, Vol.461,205–210.

Poustini K.; Siosemardeh A.; Ranjbar M. (2007) Proline accumulation as a response to salt stress in 30 wheat (*Triticum aestivum* L.) cultivars differing in salt tolerance. *Genetic Resources and Crop Evolution*, Vol.54,925-934.

Ramon, M.; Geuns, J.M.C.; Swennen, R.; Panis, B. (2002) Polyamines and fatty acids in sucrose precultured banana meristems and correlation with survival rate after cryopreservation. *Cryoletters*, Vol.23,345-352.

Ribarits, A.; Abdullaev, A.; Tashpulatov, A.; Richter, A.; Heberle-Bors, E.; Touraev, A. (2007) Two tobacco proline dehydrogenases are differentially regulated and play a role in early plant develop- ment. *Planta*, Vol.225,1313–1324.

Roach, T.; Ivanova, M.; Beckett, R.P.; Minibayeva, F.V.; Green, I.; Pritchard, H.W.; Kranner, I. (2008) An oxidative burst of superoxide in embryonic axes of recalcitrant sweet chestnut seeds as induced by excision and desiccation. *Physiologia Plantarum,* Vol.133,131-139.

Rudolph, A.S.; Crowe, J.H.; Crowe, L.M. (1986) Effects of three stabilising agents – proline, betaine and trehalose – on membrane phospholipids. *Archives of Biochemistry and Biophysics,* Vol.245: 134–143.

Sanchez, D.H.; Siahpoosh, M.R.; Roessner, U.; Udvardi, M.; Kopka, J. (2008) Plant metabolomics reveals conserved and divergent metabolic responses to salinity. *Physiologia Plantarum,* Vol.132,209-219.

Saradhi, P.P.; Alia; Arora S; Prasad, KVSK (1995) Proline accumulates in plants exposed to UV-radiation and protects them against UV induced peroxidation. *Biochemical and Biophysical Research Communications,* Vol.209,1-5.

Siripornadulsil, S.; Traina, S.; Verma, D.P.S.; (2002) Molecular mechanisms of proline-mediated tolerance to toxic heavy metals in transgenic microalgae. *The Plant Cell,* Vol.14,2837-2847.

Smirnoff, N. (1993) The role of active oxygen in the response of plants to water deficit and desiccation. *New Phytologist,* Vol.125: 27–58.

Smirnoff, N.; Cumbes. Q.J. (1989) Hydroxyl radical scavenging activity of compatible solutes. *Phytochemistry,* Vol.28:1057–1060.

Stranska, J.; Kopecny, D.; Tylichowa, M.; Snegaroff, J.; Sebela, M. (2008) Ornithine d-aminotransferase: an enzyme implicated in salt tolerance in higher plants. *Plant Signaling & Behavior,* Vol.3,929–935.

Strizhov, N.; Abraham, E.; Okresz, L.; Blickling, S.; Zilberstein, A.; Schell, J.; Koncz, C.; Szabados, L. (1997) Differential expression of two P5CS genes controlling proline accumulation during salt-stress requires ABA and is regulated by ABA1, ABI1 and AXR2 in Arabidopsis. *The Plant Journal,* Vol.12,557–569.

Swaaij, A.C.V.; Jacobsen, E.; Feenstra ,W.J. (1985) Effect of cold hardening, wilting and exogenously applied proline on leaf pro- line content and frost tolerance of several genotypes of Solanum. *Physiologia Plantarum,* Vol.64,230–236.

Swaaij, A.C.V.; Jacobsen, E.; Kiel, J.A.K.W.; Feenstra, W.J.; (1986) Selection, characterization and regeneration of hydroxyproline- resistant cell lines of *Solanum tuberosum*: tolerance to NaCl and freezing stress. *Physiologia Plantarum,* Vol.68, 359–366.

Szabados, L.; Savoure, A. (2010) Proline: a multifunctional amino acid. *Trends in Plant Science,* Vol.15,89–97

Szekely, G.; Abraham, E.; Cseplo, A.; Rigo, G.; Zsigmond, L.; Csiszar, J.; Ayaydin, F.; Strizhov, N.; Jasik, J.; Schmelzer, E.; Koncz, C.; Szabados, L. (2008) Duplicated P5CS genes of Arabidopsis play distinct roles in stress regulation and developmental control of proline biosynthesis. *The Plant Journal,* Vol.53,11–28.

Taji, T.; Ohsumi, C.; Iuchi, S.; Seki, M.; Kasuga, M.; Kobayashi, M.; Yamaguchi-Shinozaki, K.; Shinozaki, K. (2002) Important roles of drought- and cold-inducible genes for galactinol synthase in stress tolerance in *Arabidopsis thaliana. The Plant Journal,* Vol.294,417-426.

Verbruggen, N.; Hermans, C. (2008) Proline accumulation in plants: a review. *Amino Acids,* Vol.35,753–759.

Xin, Z.; Browse, J. (1998) eskimo1 mutants of Arabidopsis are constitutively freezing-tolerant. *Proceedings of National Academy of Science USA,* Vol.95, 7799–7804.

Xin, Z.; Li, P.H. (1993) Relationship between proline and abscisic acid in the induction of chilling tolerance in maize suspension- cultured cells. *Plant Physiology*, Vol.103,607–613.

Yoshiba, Y.; Kiyosue, T.; Nakashima, K.; Yamaguchi-Shinozaki, K; Shinozaki, K. (1997) Regulation of levels of proline as an osmolyte in plants under water stress. *Plant Cell and Physiology*, Vol.38,1095–1102.

Zhang, L.P.; Mehta, S.K.; Liu, Z.P.; Yang, Z.M. (2008) Copper-induced proline synthesis is associated with nitric oxide generation in *Chlamydomonas reinhardtii*. *Plant and Cell Physiology*, Vol.49,411–419.

Part 3

Equipment and Assays

11

X Ray Diffraction: An Approach to Structural Quality of Biological Preserved Tissues in Tissue Banks

H. Perez Campos et al.*
*Instituto Nacional de Donación y Trasplante (INDT), Ministerio de Salud Pública -
Fac. de Medicina,
Uruguay*

1. Introduction

The purpose of this chapter is to introduce new methods of analysis, to evaluate the final quality of human origin bio therapeutics products generated in Tissue Banks (TB), using well developed and known techniques in various fields of Physics, Chemistry and Biology as applied X – Ray diffraction (XRD), and Raman Scattering (RS).

Cryopreservation techniques are fundamental supports in the conservation procedures of biological materials in TB work. However, controversial views remain on the effects at the molecular level that cryogenic temperatures and thawing could produce on the functional structures of tissues. The same concept can be sustained to glycerolized tissue preservation. Taking into account this scope, we implemented a methodological scheme to analyze tissue specimens before and after programmed cryopreservation, or glycerolization in order to find structural differences in the basic material constitutive collagen, using the techniques formerly mentioned: diffractive and scattering.

It is noteworthy that both methods of analysis can be applied to any type of tissue preserved for the aforesaid purposes, with other conservation techniques, such as freeze drying or un programmed freezing.

2. The tissue banks and the "viability" of it's therapeutically products

The TB are technical establishments whose main institutional objectives are collection, preservation, storage, release and distribution of biological tissues for therapeutic use in transplantation medicine. These objectives is met according to scientific criteria from agreed international protocols (Spanish Association of Tissue Banks: AEBT, International Atomic Energy Agency, IAEA, European Association of Tissue Banks: EATB; American Association of Tissue Banks: AATB) and according to the legal frameworks of the different countries and

* Mc. Saldias[1], G. Sanchez[1], P. Martucci[1], Mc. Acosta[1], I. Alvarez[1], R. Faccio[2], L. Suescun[2],
M. Romero[2] and A. Mombru[2]
[1] *Instituto Nacional de Donación y Trasplante (INDT), Ministerio de Salud Pública - Fac. de Medicina, Uruguay*
[2] *Laboratorio de Cristalografía, Estado Sólido y Materiales (DETEMA) Fac. de Química, Uruguay*

their ethical rules. On the other hand, tissue banking activities are carried out following strict bio-security criteria by the selection of suitable donors, permanent quality control and continuous management of physical plant, equipment, supplies procurement procedures, and end products, which must comply criteria for "viability" therapeutic apply.

According to (Pegg 2006) this concept of "viability," applies whenever a graft of tissue obtained as final product, meets the leading natural-biological function for which it was preserved, and that is pathologically affected in the recipient. This explains why there are different procedures and methods of conservation in TB, according to the type of tissues and its expected restored function.

In some cases, we must preserve, as the main function, a biological synthesis, which requires mandatory of vitality in cellular functions (eg, parathyroid glands, or ovarian tissue). In other cases, the objective is to preserve static and mechanical functions as a segmental allograft bone support, or biodynamic behavior, as in cases of preservation of blood vessels. This implies that "viability" is not necessarily synonymous with "cell vitality" and therefore the requirement is to preserve elements of the Extra Cellular Matrix (ECM).

Hence, the importance of assessing possible changes on the components of the ECM that TB procedures may be generate on processing and stocking biological tissues.

3. The concept of functional structure of ECM

The ECM was considered early in the twentieth century as filler material and mechanical support of cell structures, which was thought as the only protagonists of tissues functionality. In the 50´s, (Grobstein, 1953) proposed that the induction in the development of a tissue depended on the presence of ECM.

In the 60´s, (Hauschka & Königsberg, 1966) establish that pig embryonic muscle cell cultures proliferate more properly in a media with presence of metabolic products from fibroblasts, and identify the collagen to induce the development of them. This protagonist role of collagen in the processes of induction and cell proliferation, is corroborated by (Meier & There, 1974), by testing the inductive capacity of collagen, for the synthesis of ECM in the corneal epithelium of pig embryos (5 days old development). (Sanders 1988) working on neural crest and sclerotome cells of early chicken cells embryos, prove that the presence of type I collagen is necessary for cellular migration and *de novo* synthesis of ECM is a prerequisite for normal cell migration and attachment in earliest stages of embryogenesis.

Additionaly, since the early 80´s (Bissell 1982), work in the field of cancer biology emphasizes the importance of "micro" immediate cellular environment and posits the hypothesis of 'Dynamic Reciprocity' by which the ECM contact trans membrane receptors, influence gene expression through signals transmitted via cytoskeleton, generating so "new" products for the ECM. Thus, the cell and the ECM, form a binomial reciprocal exchange interaction, which has vital importance in the early stages of embryonic morphogenesis and later in postnatal life, the physiological mechanisms of growth and development as well as in response to injury. (Bissell, 1982; Davis, 2010; Kelleher, 2004; Nadiarnykh, 2010; Schultz, 2005; Schwinn, 2010). The ECM consists of a complex variety of macro molecules that can be summarized schematically as follows: 1) protein collagen, 2) structural glycoproteins, 3) proteoglycans and glucosamine glycans, and 4) elastin. These complex and diverse molecular groups, organized into super families, are shown with a dynamic distribution, and functional modulated behavior, with variations between different tissues. This bio plasticity is observed even within the same tissue type, as homeostatic biochemical, and bio mechanical requirements, including interactions with various molecules: growth factors, cytokines, enzymes and other inducers synthesis products as well as lytic and degradative matrix one.

The surface receptors of the cell membrane, in close contact with this complex and dynamic molecular ECM set, interact through the cytoskeleton to the genome, which modulates the different stages of ontogeny, growth and postnatal development sequences as well as molecular structural and functional physiology and patho physiology biological requirements of tissues. (Abraham, 2007; Bowers, 2010; Worthley, 2010). Related to the own collagen structure there are ligands and functional domains that take contact with other ECM molecules (fibronectin, proteoglicans, and collagen - collagen interactions) and with the neighbor cells microenvironment (cell integrin receptors). Several poly peptid sequences, (eg.: GFPGER: glycine - phenylalanine - hyroxyproline - glycine - glutamic acid - arginine) and ligands domains (eg.: Matrix Metalloproteinase Interaction Domain or MMP ID; Colagen V Cross-link site or Col V X-link), have been identified and play an important role in regulation of migration, proliferation, adhesion and apoptosis in biology cell tissues. (Orgel, 2011; Sottile, 2007; Sweeney, 2008).

It is obvious therefore that the extra cellular medium, forms a molecular complex of plastic, in both dynamic up regulator as down regulator in constant cross talk with cell pole whose points of contact and mutual information imply the presence of binding sites at ECM structure related to cell surface receptors. Fibril collagen constitutes approximately 25% of tissues for all species of mammals and is the main component of the total molecules that make up the ECM. (Kielty, & Grant, 2002). To date, they have been described up to 29 different types of collagens with the corresponding genetic determinants. "Structural" Collagens in the ECM are called fibrils (Fibril Collagen) consisting of types I, II and III, V and XI. Type I constitute 90% of body collagen and mainly, perform mechanical resistance functions. In addition it provides three - dimensional modeling formation of tissues. An important bio molecular feature of our study is it hierarchical and sequenced shaping showing collagen. Taking collagen I model, pro collagen, amino acid primary structure have a intracellular synthesis (endoplasmic reticulum), with repeated tripeptides, design whose residues are Gly-Pro- Hyp or Gly-X- Hyp which Pro and Hyp are near to 17% and 33% respectively. Therefore 50% of the average 1000 residues of the total composition of the molecule, pro collagen, are other amino acids. Its length of 300 nm and width of 1.5 nm, is organized in a left-handed secondary structure of three amino acids per turn, with Gly residues central and peripheral Pro and Hyp out of the spiral. Three assembly helical pro collagen monomers (2 α1 and 1 α2) in right-handed configuration, determine the tropo collagen structure in the extra cellular space. In this space the molecule is arranged in staggered bundles with a gap of 67 nm by inter molecular bonds tropo collagen units, which gives to the collagen fibril new product design, a repetitive sequence which observed in the ME identified a characteristic D - banding. The final design shows collagen fibers arranged, spatially distributed in regular packages along the lines of force of the biomechanical characteristics of each tissue. This last aspect brings an added dimension of ordering design given by the spatial distribution of fiber bundles, and their inter reciprocal space.

These three characteristics: a) repetitive and periodic sequencing of the D - banding, b) structuring hierarchically ordered by supra molecular complexes, and c) nano-scale dimensions of the structures, made of collagen complex an para crystalline super molecule, liable to be analyzed by techniques diffractive as discussed later. (Sweeney, 2008; Berenger, 2009).

In this context becomes important the analysis of the changes that for preservation purposes, can be induced in the molecular components of the ECM. Particularly taking account that collagen is the main structural component of the extra cellular microenvironment and has a proven role in the functional biological mechanisms, developmental, physiological homeostatic and physio - pathological tissues behavior. (Kielty & Grant, 2002; Orgel, 2011).

These reasons justify work TB, to design tissue preservation processing models, with conservation of the collagen component, from both, structural and biochemical characteristics. It must be take in mind, that allografts should meet a homeostatic interaction with cell biology recipient patient, through its membrane receptors in functional contacts with molecular ligands and domains of preserved ECM, to improve physio pathological situations.

Note therefore that the biological behavior of a suitable allograft depend on the presence and indemnity of molecular epitopes or ligands, which can be eventually altered in its stereo chemical distribution during cryogenic or glycerolized procedures.

4. The interaction between ECM components, and the physic-chemical phenomena preservation procedures

About the cryogenic effects on ECM, there are several references in different disciplines, about the ultra cold temperatures on biological material. Indeed, at the molecular level have been observed different types alterations generated by freezing / thawing phenomenon, on biological structures. As earlier in the 60´s, (Levit, 1962; 1966), had postulated irreversible changes in the tertiary structures of soluble vegetal proteins, with loss of its biological capacity, such as the rearrangement of disulfidric functional bonds to non-functional disulfide covalent configurations.

The concept of "repulsion hydration forces" refers to cell membranes, was developed by Wolfe, J. (1999). These phenomena is induced by water efflux through such semi-permeable membranes during cooling process, promoting large mechanical stress and strain in the biological structures, and generating physical deformations and changes in the molecular functional membrane behavior. Such condition is done under the observed ground of the structure of the crystalline ice mass in the extra cellular space. This, results in intra cellular dehydration of the tissue and the extra cellular hyper osmolarity of super cooling liquid. So, it generates a displacement of inter atomic and molecular chemical equilibrium that, explains changes in the stereochemistry and molecular architecture of biological structures. This scope would agree to Levit postulates. (Levit, 1962; 1966).

The effect that the conventional criopreservation exerts on the ECM structure is controversial information. (Gerson, 2009), comparing morpho structural collagen mesh from fresh and cryopreserved human heart valves by second harmonic generation, sets no changes between both categories. However, it was found extensive damage in collagen structure in porcine frozen leaflets related to fresh control one, using laser-induced auto fluorescence imaging , (Schenke Layland, 2006), and second-harmonic generation. (Schenke Layland, 2007).

In other field, there are many studies showing that the biomechanical behavior of tissue collagen framework, is not altered by effect of cryopreservation / thaw cycle, in vascular (Armentano, 2006; Bia, 2006; Langerak, 2001, 2007; Pukacki, 2000), tendon , (Woo, 1986; Park, 2009) or bone tissues. (Hamer, 1996).

However, controversial literature is also observed for biomechanical variables. (Rosset, 1996) observed in vitro, decreased compliance and hysteresis an increase of modulus of elasticity in thawed cryopreserved human carotid arteries, related to fresh one. (Gianni, 2008) found that the freezing of human posterior tibial tendons significantly affected behavior in vitro biomechanical performance. Finally, either way, under many point of view is possible highlight that the functional character of fibril collagen depending to the particular structural and biochemical preservation, which may be damaged during cryopreservation defrosted process in TB.

About the interaction between alcohols and polymerized amino acid, in early 70´s (Frushour, & Koenig, 1975) postulated, in Raman Scattering field, that methanol modified an aqueous poly-DL-alanine (PDLA) solution, by disruptions of the helical regions by breaking the hydrophobic bonds.

5. Diffractometry: a tool for analysis of structural ordering of collagen

5.1 The matter and its organization

It defines that the spatial arrangement of ions and atoms of matter, have a crystalline profile when its design shows a repeating sequence. The frequency of repeated and symmetrical distribution of the atomic stereo chemical units constituents of matter, determine the solid crystalline character. It is understood that a substance is "homogeneous" when each constituent unit of the solid is linked by chemical bonds to another identical unit in any sense of space, and is identified as the ideal "homogeneous crystal" model when, theoretically, is infinitely extended in space. The sequential nature of the repetitive and symmetric atomic elements, defines the spatial network model, under the so called "cells ordering". The three dimensional symmetric distribution of elemental units of the complex let likened to an orderly succession of planes separated by a distance "d". This design is easily identifiable in crystalline substances of inorganic chemistry: quartz, diamond, graphite, etc. Diffracted analysis of inorganic or organic crystalline matter can provide detailed information about molecular design, related to intermolecular distance and stereo chemical angle conformation. In the world of bio molecular chemistry, matter is organized by more complex models, through an extensive variety of atomic molecular combined structures. In this picture certain combinations become repetitive units, consisting of several basic types of atoms links together by different kinds of bonds. Nevertheless, one can observe the character of certain spatially ordered molecular configurations, which are equally repetitive. This setting defines the so called "molecular crystals", despite not showing the perfection system of "homogeneous crystals". So, no bond lengths and angular atomic positions can be determined, but an approximate view about relative ordering structure is given applying XRD techniques. These structures are typical of biological substances such as proteins or DNA molecules whose functional design depends on the molecular arrangement, and type of chemical bond established between its molecules.

5.2 The diffractive phenomenon and elastic scattering of x-rays

X-rays are a form of electromagnetic energy produced by a source to be impacted by electrons of high kinetic energy supplied (usually a tungsten filament named cathode). The incident electron impact, destabilizes the internal atomic orbital of a target material (an anode built with pure copper), generating atoms in electronically excited state. The movement of electrons from outer orbital to balance the impact generates heat energy and emission of x-rays, in a spectrum of wavelengths (λ) measurable in Angstrom units ($1 \text{ Å} = 10^{-10} \text{ m}$). The band spectral x-rays emission is "filtered" through mono chromator to obtain a single wavelength that corresponds to the "characteristic radiation" used in the x-rays crystallography: K-alpha line (Kα) for each material property anode of the device. To our work, XRD is defined by the interference between monochromatic characteristic emissions (Kα) with the ordered material in crystalline form. When a photon interferes with an orderly molecular structure without loss of energy (elastic scattering), produces a deviation from its original direction which is the diffraction phenomenon. The condition for diffraction to be possible is that the distance between the periodic and ordered structure elements (atoms or molecules), fall in the wavelength range of incident ray. The condition to be detectable is that some degree of ordering is present in the material to be studied, in order that the interference could be constructive. As previously mentioned, collagen shows particular characteristics like a biomolecular arrangement built by: a) the hierarchical supra molecular order fibers, b) the observed crystalline structure of the spaces sequenced "D" of fibrillar collagen, c) the nano scale distance of repetitive molecular units, all of them allows to analyze biological stroma collagen by the XRD. (Aspden, 1987; Berenguer, 2009; Connon, 2007; Hickey & Hukins, 1980; Horton, 1958; Pauling & Corey, 1951; Pérez Campos, 2008).

The equation that allows the practical application of this technique is Bragg's Law.

$$n \lambda = 2d \sin \theta \qquad (1)$$

Formula 1 Bragg´s Law: n is an integer, λ is the x – ray monochromatic wavelength, d is de distance between the planes of the crystalline net, and θ is the angular value between the incidental x –ray and the considered crystalline plane.

Given a certain incidence angle of a monochromatic beam on the material structure, sequencing crystalline spacing "d" determines a dispersive interference when the rays are emerging in construction phase. Emerging rays can be recorded in a Cartesian coordinate system where the independent variable "x" records the range of 2θ values and the dependent variable "y", the relative values (R.V.) of the ordering lattice system. The recordable diffraction graphics or diffractogram depend of the content and the atomic distribution within the repetitive units that define the three dimensional arrangement. (See Figure 1)

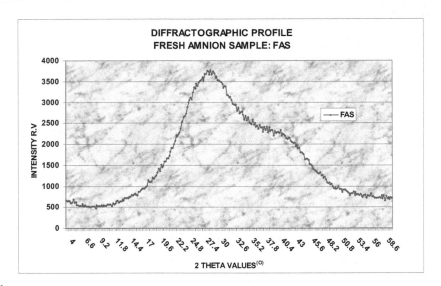

Fig. 1.

5.3 The Raman effect, or inelastic scattering of light

Given a monochromatic light beam incident on a material, there will be a phenomenon of elastic light scattering as a result of the interaction of photons and electron atomic elements of the network links. The elastic scattering implies that the frequency of the incident light beam and the scattered light emerging is the same, so that no changes have occurred in the respective energy levels. This phenomenon known as Rayleigh scattering is highly significant from a statistical point of view. However an extremely low intensity incident light (in the order of 1 photon in 10^7 to 10^{10}) shows inelastic behavior in the interaction with the structure determining slight changes in the emerging wave frequency, which depends on the characteristics of matter incised. This phenomenon or Raman Effect was discovered by (Raman & Krishnan, 1928), and allows the chemical structure analysis of biological material.

Atomic particles mass and its energy states (vibration and/or rotational), maintain chemical bonds that define crystal structure, sustaining dynamic design stability by neighbor interaction. The movements of both vibration and rotation of the particulates: (υ = frequency), defined the dynamically stable energy level where they are. If the interaction of a beam of photons of frequency (υ_0), print a change of unstable frequency at the particle network link, they scatter photons with a different frequency (υ_r), define an inelastic scattering. The energy can be dispersed in a model $\upsilon_r > \upsilon_0$ called scattering Raman - Stokes or $\upsilon r < \upsilon_0$ know as scattering Raman anti Stokes. υ_r values are characteristic of each design structures and define atomic molecular matter. Approximately 99% of the output assays is Stokes Raman scattering hence, those are the models profiles recorded. In the Cartesian co ordinate system the independent variable x records the difference υ_r-υ_0 in cm^{-1}, and the ordinates y, the scattering intensities for each differential rate, in relative units (R.U.) (See Figure 2)

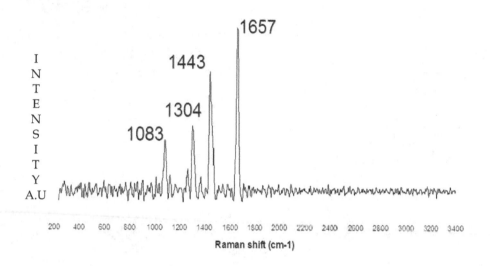

Fig. 2. Raman Spectum oleic acid

5.4 Interaction x-rays - Collagen: A model analysis

The x-rays diffractive analysis of the collagen structure was earlier studied (Pauling & Corey, 1951). Defined diffraction peak was described, in the range d = 2.86 Å (2 θ = 31.3 ° using CuK α radiation λ = 1.5418 Å value), based on the criteria of the Bragg Law. This phenomenon was interpreted as the expression of cis configurations for the amide groups of the polypeptide chain.

Using this background, our group analyzed tissue banking allograft, in order to compare the diffraction profiles obtained before and after cryo-preservation method (vascular tissues and amnion tissue), and glycerolized preserved method (amniotic membrane). The working hypothesis states that the preservation methods can modify the stereochemistry molecular structures, determining changes in collagen and the consequent differentiation of diffractive or dispersive profile.

6. Materials and methods

6.1 Donors and biological samples procedures

The applied procurement protocol to vascular tissues was made on cadaver multi organ donors through informed consent and in accordance with standard operating manuals in the National Institute for Donation and Transplantation (INDT) of Uruguay. They were likewise applied legal and ethics regulations (Law 14005/1971 - 17668/2003) valid in our country. The exclusion criteria and biological safety were applied, in accordance with the Standards for Tissue Banking: International Atomic EnergyAgency (IAEA - 2005) and the

Spanish Association of Tissue Banks (AEBT - 2005). The same selection criteria, exclusion, tissue procurement and processing were applied to living donor placenta with clinical controlled normal pregnancy and delivery, by the parameters set by the Ministry of Public Health of Uruguay.

Donors were selected according to the protocol in a range between 18 and 60 years (35.5 ± 11.8 years mean age, 47% M 53% F) obtained by aseptic dissection, 10 aortic arterial segments, and 8 carotid. It proceeded under a laminar flow cabinet to cleaning, package, and storage in physiological saline at 4°C. Fresh Vascular Samples (FVS) were shipped to DETEMA within 24 hs, for XRD. The same process aseptic protocol was applied to 6 amnion obtained by manual dissection from donor placenta. Fresh Amnion Samples, (FAS) were stored in saline solution at 4°C and shipping to DETEMA within 24 hs for XRD analysis.

The Cryopreserved Vascular Samples (CVS), segments of each contra lateral carotid donor, and hemi ring segments of thoracic descendent aorta, were processed for cryopreservation in a Controlled Rate Freezing System (Model 9000, Gordinier Electronics, Inc. Michigan). Stored CVS were maintained up to 30 days at -142°C. The cryopreservation media was: RPMI 1640, 85 cc; Human Albumin (20%), 5 cc; DMSO 10%. Cryopreservation was made into termal sealed double cryo resistant bag (Joisten and Kettenbaum D51429, Bereisch Gladbach, Mod.011342). The mean cooling rate applied was -1°C/min from 4°C to -90°C and then quickly stored at -142°C during 30 days in steam liquid nitrogen.

Same procedures were applied to obtain 2 Cryopreserved Amnion Samples (CAS), stored up to 30 days at -142°C.

Defrost protocol applied for vascular and amniotic tissues, was according with Pegg et al. (1997), and defrosted samples were shipped to DETEMA for XRD.

6 Glycerolized Amnion Samples (GAS) were obtained by soaked in screw cap flask in Glycerol (95%) and stored at 4° C for 30 days. Glycerol from amnion was removed by three sequential shaking washing for 15 min. each, in saline solution and then shipped to DETEMA for x-rays diffraction, and Raman Scattering analysis. The comparative assays were done with FVS vs CVS; FAS vs CAS; and FAS vs GAS.

6.2 Diffractographic and Raman scattering technical procedures

XRD measurements were conducted at the Laboratory of Crystallography, Solid State and Materials, School of Chemistry (DETEMA), with a CuKα radiation source of wavelength λ = 1.5418 Å, using a Rigaku Ultima IV diffraction system. The incident ray is calibrated to arterial vessels in a range for 2θ between 5 ° and 60 °, step scan of 0.1 ° for 10 sec. each. The respective diffraction profiles (FVS vs. CVS) were filed for later analysis. Comparative profiles for amnion (FAS vs. GAS; and FAS vs CAS) were treated the same way as having been calibrated for 2θ between 5° and 60 ° ranges scanning with steps of 0.2° for 10 sec each.

Raman spectra were recorded using a Raman DeltaNu Advance 532 spectrometer with a laser frequency doubled Nd: YAG, 100mW, with a 532 nm wavelength, scanning in the 200 and 3400 cm^{-1} region.

6.3 Planimetric analysis: Obtaining the order coefficients for XRD (Perez Campos, 2008)

Given the diffraction profiles of two tissues A and B to compare tests, we can define the respective planimetric surfaces, defined under the corresponding diffraction curve, that are a function of the degree of molecular arrangement of the studied tissue.

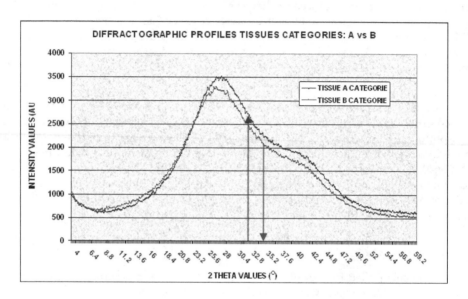

Fig. 3. Diffractive profiles from two different tissues categories A vs B.

Relative Differential Intensity values (RDIV) established for each 2θ point between $5°$ and $60°$, will produce a result that: if absolute value tissue A > tissue B will have a resulting positive (+) value; but if tissue A < tissue B will have a resulting negative (-) value.

With those values obtained in each point from 2θ it can be developed Differential Planimetric Surfaces (DPS) that represent the sum of every relative intensity value.

Its mathematical expression is given by the equation:

$$\Sigma\ (\ \uparrow - \downarrow\ )\ \times I_{A-B}\ [2\theta\ (°)] = DPS \qquad (2)$$

Where ($\uparrow$) represents ordering diffractometric intensity for y axes values of tissue A in one point in 2θ; ($\downarrow$) represents ordering diffractometric intensity for y axes values of tissue B, in the same point in 2θ; I_{A-B} is the difference between each respective ordering diffractometric intensity for y axes value at the same point in 2θ. Finally, $[2\theta\ (°)]$ is each point in x axes between $5°$ and $60°$ angular incidence.

DPS can be edited in a Cartesian model too, where x axis is 2θ values and y axes is the Intensity Relative Differential Values (IRDV) between both comparative samples for each point 2θ values. (See figure 4)

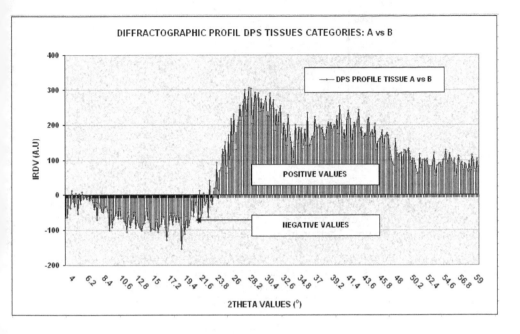

Fig. 4. DPS from two tissues categories A vs B to be studied by X-ray diffraction.

The ratio DPS (+) values vs DPS (-) values define the Ordering Profile Coefficient (OPC) according to the following formula:

$$OPC = \frac{DPS + VALUES}{DPS - VALUES} \qquad (3)$$

OPC absolute values are always above 0 and they are greater than 1 when +DPS values > -DPS values. When +DPS values < -DPS, OPC falls into an interval greater than 0 and lower of 1.

7. Results

7.1 Cryopreserved vascular tissues results

Analysis of the diffraction curves shows that regardless of the condition FVS or CVS, the same design with a peak of maximum intensity to 31.3 ° and another lower, at 42 ° in 2θ is kept, whether there are noticeable differences in design profiles between the two categories, even for a single donor. (Perez Campos, 2008). Comparatives diffraction profiles shows the confirmation of a peak intensity for 2θ = 31.3 ° corresponding a d - spacing = 2.86 Å. The lower peak intensity, obtain d – spacing = 2.15 Å applying Bragg Low calculus. This behavior is independent of the vessel (aorta or carotid) and the processed sample (FVS or CVS). See Figure 5

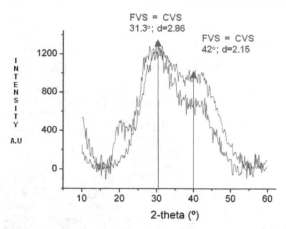

Fig. 5. Diffractografic profiles from thoracic descending aorta. Code color: FVS, Red; CVS, Blue. Note: The 1st and 2nd maximum peak labels of each respective diffractive curve indicate; Category of tissue: 2 Theta value; and calculated d spacing.

Calculated OPC values in respective analyzed vessels, shows a greater crystalline framework for CVS vs FVS, regardless the kind of arterial segment: aorta or carotid. 75% of aortic samples showed OPC values > 1 and 62,5% of carotid samples had the same behavior. Perez Campos et al (2008). Figure 6 show DPS defined from diffractografic FVS and CVS of Figure 5:

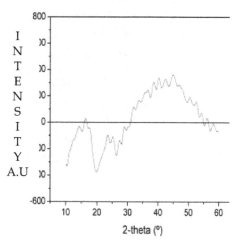

Fig. 6. DPS profile FVS vs CVS in a descending thoracic aorta from a male donor 50 years old.

Note the great difference of design shape of DPS curve related to the same one obtained from amnion tissues; (see below).

7.2 Glycerolized amnion tissues results

The diffraction curves of the glycerolized amniotic membrane, also shows the same kind of form and design for both the FAS and GAS. Notwithstanding the maximum diffractive peak in FAS d spacing = 3.24 (28.4 ° in 2θ) and in GAS, d spacing = 3.28 (28 ° in 2θ). A second peak is shown to both kinds of samples FAS and GAS for same d spacing = 2.35 (40.4° in 2θ). Contrary to the notable profiles differences showing in the two categories of vascular tissues (FVS and CVS), both amnion profiles -fresh and glycerolized- have almost the same design curve. (See Figure 7)

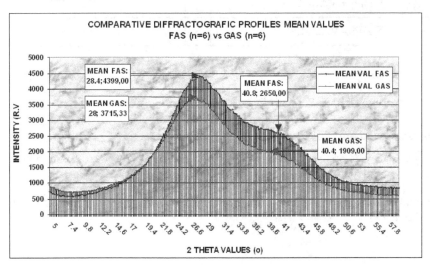

Fig. 7. Diffractive curves FAS and GAS profiles. Note: The 1st and 2nd maximun peak labels of each respective diffractive curve indicate; Categories of tissue: 2 Theta value; and Intensity RV.

Mean diffractographic profiles for 6 FAS vs 6 GAS let us obtain DPS picture and calculate OPC values = 14.76 (See respective Figure: 7 and Table: 1)

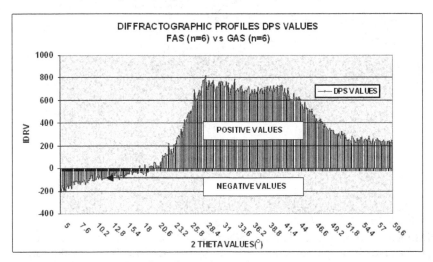

Fig. 8. DPS profile by FAS vs GAS analysis obtained.

OPC CALCULATION FAS vs GAS		
OPERATION		ABS. VALUES
$\sum$ DPS + VALUES		94228,17
$\sum$ DPS - VALUES		-6385,33
OPC VALUE = (+DPS) / (-DPS) = 14,76		

Table 1. OPC value from FAS vs GAS.

7.3 Cryopreserved amnion tissues results: X-ray diffraction

4 FAS vs 2 CAS was analysed. The same phenomena of Glycerolized amnion about form and design maintenance, was observed between FAS and CAS. Equally, there are a maximum peak in FAS d spacing = 3.26 (28.2° in 2θ) and in CAS, the same d spacing. Also a second peak is detectable to FAS in d spacing = 2.36 (40.8° in 2θ) and to CAS with equal values. (See Fig 9)

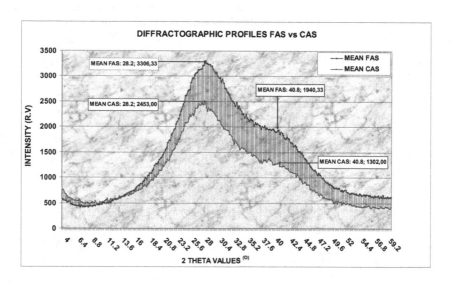

Fig. 9. Diffractive curves FAS and GAS profiles. Note: The 1st and 2nd maximun peak labels of each respective diffractive curve indicate; Categories of tissue: 2 Theta value; and Intensity RV.

Figure 10: show DPS profile obtained from operative subtractions analysis between FAS Vs CAS respective diffractive curves:

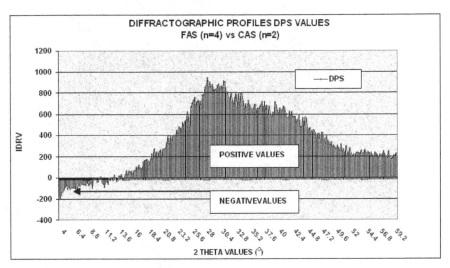

Fig. 10. DPS planimetric picture FAS vs CAS.

Operative planimetric values DPS obtained showed an OPC = 42.02 (See Table 2).

Newly highlight the notable differences in shape and form between vascular and amnion DPS pictures.

OPC CALCULATION FAS vs GAS		
OPERATION		ABS. VALUES
$\sum$ DPS + VALUES		102887.67
$\sum$ DPS - VALUES		-2448,33
OPC VALUE = (+DPS) / (-DPS) = 42.02		

Table 2. OPC values FAS vs GAS

7.4 Cryopreserved amnion tissues results: Raman Spectra

Figure 11 Show Raman Spectra profile from FAS vs CAS assays. Note that having regard to the best imaging definition, and taking account the meaningful change area of Raman Shift between both categories, the showed range is 1000 – 2000 cm^{-1} in x.

It should be highlighted that arrows points marked correspond to noticeable differences in positive Intensity values (AU) between FAS (red line) and CAS (blue line).

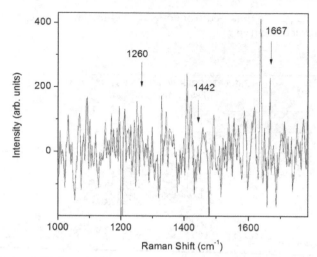

Fig. 11. Raman Spectra FAS vs CAS to area 1000 – 1800 Raman Shift (cm-1). Color code; Red: FAS, Blue: CAS.

8. Comments

8.1 X-ray diffraction

The application of XRD on final quality of stromal collagen tissues analyzed, show differential results according tissue type and / or method of preservation applied. Indeed, relative to the observed results in cryopreservation of arterial vessels we see that regardless of the vessel –carotid or descending thoracic aorta- and the condition of FVA or CVS, a common diffractive peak at d spacing 2.86 Å is seen. The same phenomenon is shown for a second peak at d spacing 2.15 Å. These results show that vascular cryopreservation -defrost procedures did not alter the sequential structure of vascular fresh collagen. In reference to the results of amniotic membrane processing under the same preservation procedures, we see that for both varieties, fresh and cryopreserved, the maximum diffractive profiles remain unchanged: FAS, with d spacing 3.262 Å (28.2 ° in 2 θ) and equal values are checked for CAS. Other 2nd peak at d – spacing 2.359 (40.8° in 2 θ) is verified for both study categories. This confirms that the collagen is resistant to the cryopreservation defrost technique in regard to its sequential molecular structure independently of type of tissue.

Contrary, when both amniotic membrane categories study values are observed we found that FAS variant shows a peak for d spacing of 3.241 Å. (28.4 ° in 2θ), while GAS is expressed in the maximum deflection for spacing d 3.284 (28 ° in 2θ). This lag is not verified for the 2nd peak in both categories that match at d – spacing = 2.359 (40.8° in 2 θ). (See figure 7). These findings showed significant data in the sense that the chemical preservation of amniotic membrane with glycerol, modifies sequencing molecular design of collagen, while this variable is not changed under the cryopreserved defrosted condition.

These would be in according with aforementioned work from (Frushour & Koenig 1975).

Additionally, by analyzing the profiles of ordering by OPC values -in relative terms- we see that both, the cryopreservation defrosted and glicerolización procedures, down modify profiles designs, defined as "molecular crystals". (FAS vs GAS, OPC value = 14.76; FAS vs. CAS, OPC value = 42.02 respectively).

The main conclusion from these data is: amnion chemical glycerolized procedures, change sequencing molecular design, while physical cryopreserved method does not. But, physical cryopreservation method, and chemical glycerolized, modifies OPC values related to both tissue categories: cryopreserved amnion and vascular tissue.

It must be noted the observed differences in the profiles of diffractive curves between fresh and cryopreserved arterial vessels. Indeed, there are a disparity between those varieties, which were not verified by the corresponding samples fresh and cryopreserved amniotic membrane, that maintain substantially similar profiles. This aspect is independent of the OPC values for both tissues and categories of each study. Our hypothesis is that these differences are related to the anatomical and functional collagen distribution in different tissues. Indeed, the hierarchical order of collagen mesh reach a final design bundles arranged following the lines of force according to bio mechanical requirements.

In this sense, arterial wall of large conduit vessels such as aorta and carotid, are under pulsate hemodynamic regimens alternating expansion and elastic contraction states. Under these conditions, the main loads acting on the vessel wall are pressure and blood flow. The blood pressure acts directly on the inner wall of the vessel in normal direction, and flow proactively work generating a pressure proportional to the square of blood velocity, Fung YC (1997). There are therefore two preferred directions in the distribution of the charges: one circumferential and other longitudinal. This results in a complex morphological organization of the cellular components of the middle layer, composed of smooth muscle cells and collagen mesh ECM, whose design will trace circumferential and longitudinal lines, giving to vascular tissue an anisotropic mechanical behavior condition. (Rodriguez, 2007). The primary ice spontaneous nucleation happen at random in many sites of ECM, Muldrew. 1999). The crystallization growth front, follows the preferred direction lines according to design collagen mesh. According with OPC values, the new organizational picture of the cryopreserved defrosted vascular tissues would be the result of a complex sequence of physic chemical events through preservation procedures, applying changing a complex vascular structural tissue.

This is not the organizational situation of amnion collagen ECM. Amnion membrane has a laminar design, and is not subject to biomechanical pulsate regimen. Its function as an external fetal covering meets fundamentally amniotic liquid metabolic regulations, more than biomechanical functions. In spite of, both kind of tissue studied, (vascular, and amnion membrane) have basically the same fibril collagen composition, namely: Collagen I, III, and V. Additionally, Colagen IV in structural Basements Membranes.

Then, our hypothesis is that the morphologic compositions, and the architectural organizations, according to specific functional requirements to each tissue, define the proper molecular assembly, and therefore its own diffractografic profiles.

8.2 Raman scattering

Our preliminary results obtained on amnion membrane (FAM vs CAS) show punctual differences between both categories in three ranges of Raman Sepectra: 1260, 1442, and 1667 cm^{-1} band (See figure 11) where is observed an increased Intensity (AU) values in FAS related to CAS. According to references Frank, C. et al (1995), these Raman Spectra range areas aforementioned belong to fibril collagen I, III and V from human placenta. The defined corresponding chemical residues assignments, (Frushour & Koenig 1975). are respectively: Amide III; CH3, CH2 (deform); and Amide I for each range Raman Spectra recorded.

These findings support our work hypothesis about the potential power of cryopreservation procedure to change collagen structure at molecular level.

9. Conclusions

The aim of our study was the analysis of tissues produced in TB for therapeutic purposes, by known techniques, XRD and RS, able to approach the study of structures at the molecular level, mainly in reference to collagen, the fundamental component of ECM,. Taking account the hypothesis that preservation techniques introduce changes in the matrix elements of the tissues, we subjected amniotic membrane and arterial vascular samples under two types of procedures: a) cryopreservation-defrost as physical process, and glycerolization - deglycerolization as chemical process. Regarding to the results of our tests, we accept our hypothesis and concluded that while cryopreservation modifies the structural arrangement of collagen at the level of ECM, glycerolization changes molecular d spacing of biological polymers, besides the aforementioned order. However, the changes between the processed vascular tissues and amniotic membrane are different, because while the vascular cryopreservation increases the molecular order of the crystalline structure measured by OPC value, the amniotic membrane glycerolization and cryopreservation decrease the referred molecular order. These differences are interpreted as the result of complex physicochemical phenomena that occur during preservation procedures on molecular structures and its designs. These phenomena promote variations in the tissue molecular complexity and order distribution. Preliminary data from the Raman tests corroborate the hypothesis of specific modifications in the molecular structures. The consequences of these findings on the allograft biological behavior applied to clinical purposes is a challenge to research and development. Both types of tissues studied are widely applied in the world with beneficial results for the restoration of altered structures and functions in the recipients. But the gold standard allograft is not yet produced, so it is necessary to obtain the allograft that better reproduce the structural and functional conformation of the patient implanted. This objective will be achieved through the best inter relation between recipient structures and the preserved tissues applied, at molecular level to obtain the better possible allograft behavior and patency. Mainly, taking account the advanced development of applied bio engineering, and the design of complex products (composites) that combine different types of cells and artificial, biological, or modified scaffolds.

10. References

Abraham LC, Dice JF, Finn PF, Mesires NT, Lee K, & Kaplan DL., (2007) Extracellular matrix remodeling--methods to quantify cell-matrix interactions. Biomaterials.. 28(2):151-61. PMID: 16893566

Armentano RL, Santana DB, Cabrera Fischer EI, Graf S, Perez Cámpos H, Germán YZ, Carmen Saldías MD, Alvarez I. (2006). An in vitro study of cryopreserved and fresh human arteries: a comparison with ePTFE prostheses and human arteries studied non-invasively in vivo. Cryobiology. Feb;52(1):17-26. ISSN 0011-2240

Aspden, RM., Bornstein NH., & Hukins DW., (1987) Collagen organization in the interspinous ligament and its relationship to tissue function. *J. Anat.*. 155: 141 -151. PMCID: PMC1261882

Berenguer F., Wenger MPE., Bean RJ., Boze L., Horton MA., & Robinson IK. (2009).Coherent X-ray diffraction from collagenous soft tissues. *PNAS*. (106) 15297 – 15301. PNAS: 0905151106

Bia D., Pessana F., Armentano R., Pérez Campos H., Graf S., Zócalo Y., Saldías M., Perez N., Alvarez O., Silva W., Machin D., Sueta P., Ferrin S., Acosta M.,& Alvarez I. (2006). Cryopreservation procedure does not modify human carotid homografts mechanical properties: an isobaric and dynamic analysis. *Cell Tissue Bank*. 7(3):183-94. PMID:16933040

Bia D., Armentano RL., Zócalo Y., Pérez Cámpos H., Cabrera FE., Graf S., Saldías M., Silva W., & Alvarez I. (2007). Functional properties of fresh and cryopreserved carotid and femoral arteries, and of venous and synthetic grafts: comparison with arteries from normotensive and hypertensive patients. *Cell Tissue Bank.*; 8(1):43-57. PMID:16826454

Bissell MJ., Hall HG., & Parry G. (1982). How does the extracellular matrix direct gene expression? *J. Theor. Biol.* 99:31–68 PMID: 6892044

Bowers SL., Banerjee I., & Baudino TA. (2010). The extracellular matrix: at the center of it all. J Mol Cell Cardiol. Mar;48(3):474-82. PMID:19729019

Brewster L., Brey EM., & Greisler HP. (2007). Blood Vessels. In: *Principles of Tissue Engineering.* R. Lanza, R. Langer and J. Vacanti Ed. 568 – 584. ISBN: 978-0-12-370615-7

Davis GE. (2010). The development of the vasculature and its extracellular matrix: a gradual process defined by sequential cellular and matrix remodeling events. *Am J Physiol Heart Circ Physiol.* 299 (2): H245-247. ISSN: 0363-6135

Frushour, B. G. and Koenig, J. L. (1975), Raman spectra of D and L amino acid copolymers. Poly-DL-alanine, poly-DL-leucine, and poly-DL-lysine. *Biopolymers*, 14: 363–377. doi: 10.1002/bip.1975.360140210

Fung, Y.C. (1997). Blood flow in arteries. In: Boold Flow in Arteries. *Biomechanics Circulation.*. Springer-Verlag Ed. N. York. Pp.: 108 – 200.- ISBN: 0-387-94386-6

Grobstein, C. (1953) Morphogenetic Interaction between Embryonic Mouse Tissues separated by a Membrane Filter. *Nature.* (172): 869 – 871. ISSN: 0028-0836

Kelleher CM., McLean SE.,& Mecham RP. (2004) Vascular Extra cellular Matrix and Aortic Development. Current Topics in Developmental Biology. 62: 153-188. PMID:15522742

Gerson C.J., Goldstein S., & Heacox, A.E. (2009) Retained structural integrity of collagen and elastin within cryopreserved human heart valve tissue as detected by two-photon laser scanning confocal microscopy. Cryobiology. 59(2):171-179. PMID:19591817

Giannini S, Buda R, Di Caprio F., Agati P., Bigi A., De Pasquale V. & Ruggeri A.. (2008). Effect of freezing on the biomechanical and structural properties of human posterior tibial tendons. *Int Orthop* 32(2):145–151. PMCID: PMC2269010

Hamer, AJ., Strachan JR., Black MM., Ibboston CJ. Stockley I., & Elson RA. (1996). Biomechanical properties of cortical allograft bone using a new method of bone strength measurement: a comparison of fresh, fresh-frozen and irradiated bone. J Bone Joint Surg Br. 78(3): 363 – 368. PMID: 8636167

Hauschka S. & Konigsberg I.R. (1966) The influence of collagen on the development of muscle clones. *Zoology;* 55: 119 – 126. PMCID: PMC285764

Hickey, DS. And Hukins DWL. (1980) X-ray diffraction studies of the arrangement of collagenous fibres in human fetal intervertebral disc. *J. Anat.* 131 (1): 81 – 90. PMCID: PMC1233288

Horton, WG. (1958) Further observations on the elastic mechanism of the inter vertebral disc. *J. Bone and Joint Surg.* 40 (3) 552 – 558. PMID: 13575471

Kielty, C. M., & Grant, M. E. (2002). The collagen family: structure, assembly, and organization in the extracellular matrix. In *"Connective Tissue and Its Heritable Disorders: Molecular, Genetic, and Medical Aspects"* (P. M. Royce and B. Steinmann, eds.), pp. 159–221. Wiley-Liss, New York. ISBN: 9780471251859

Langerak S. E. , Groenink M., van der Wall E. E., Wassenaar C., Vanbavel E., van Baal M. C. & Spaan J.A. E. (2001). Impact of current cryopreservation procedures on mechanical and functional properties of human aortic homografts. *Transpl Int.* 14(4):248-255. PMID:11512058

Levit, J. (1962). A sulfhydryl-disulfide Hipótesis of Frost Injury and Resístanse in Plants. *J. Theoret. Biol..* 3: 355 – 391.-

Levit, J. (1966). Cryochemistry of Plant Tissue. Protein Interactions. *Cryobiology.* 3. 243 – 251. PMID: 5970348

Meier S. & Hay E. (1974). Control of corneal differentiation by extracellular materials. Collagen as a promoter and stabilizer of epithelial stroma production. Developmental Biology. Volume 38, (2): 249-270 PMID: 4275424

Muldrew, & K., McGann, L. (1999). Cryobiology - A Short Course In: *http://www.ucalgary.ca/~kmuldrew/cryo_course/cryo_*

Nadiarnykh O., LaComb R. B., Brewer M. A., & Campagnola P. J. (2010). Alterations of the extracellular matrix in ovarian cancer studied by Second Harmonic Generation imaging microscopy. BMC Cancer, 10:94. In: *www.ncbi.nlm.nih.gov/pmc/articles/PMC2841668/* PMCID: PMC2841668

Orgel J.P., San Antonio J.D., & Antipova O. (2011) Molecular and structural mapping of collagen fibril interactions. *Connective Tissue Research.* 52 (1): 2-17 PMID: 21182410

Park H. J., Urabe K., Naruse K., Onuma K., Nemoto N. & Itoman M. (2009).The effect of cryopreservation or heating on the mechanical properties and histomorphology of rat bone-patellar tendon-bone. *Cell Tissue Bank.* 10:11–18 PMID: 18830689.

Pauling, L & Corey, R. B. (1951) The structure of fibrous proteins of the collagen gelatin group. *Proc. Nat.Acad. Sci.* 37:272 – 281. ISSN 0027-8424

Pegg D.E., Wuseman M.C., & Boylan S. (1997). Fractures in Cryopreserved Elastic Arteries *Cryobiology* 34, 183–192. PMID: 9130389

Pegg, D. The preservation of tissues for transplantation. 2006. *Cell and Tissue Banking.* 7: 349 – 358. PMID: 16957871

Perez Campos, H., Saldias, MC., Silva, W., Machin, D., Suescun, L., Faccio, R., Mombru, A., & Alvarez I. (2008). Control of Cryopreservation Procedures on Blood Vessels

Using Fiber X-Ray Diffraction. *Transplantation Proceedings*. 40(3):668-674.PMID:18454983

Pukacki F., Jankowski T., Gabriel M., Oszkinis G., Krasinski Z., & Zapalski S. The mechanical properties of fresh and cryopreserved arterial homografts. *Eur J Vasc Endovasc Surg*. 2000. 20 (1): 21-24. PMID: 10906292

Raman C.V. & Krishnan, KS. A New Type of Secondary Radiation. *Nature*. 1928. *121(3048)*, 501 - 502. ISSN: 0028-0836

Rodriguez, J. Goicolea J. M., & Gabaldón F. (2007). A volumetric model for growth of arterial walls, with arbitrary geometry and loads. *J. Biomechanics*. 40: 961 - 971. PMID: 16797020

Rosset E., Friggi A., Novakovitch G., Rolland P.H., Rieu R., Pellissier J.F., Magnan P.E. & Branchereau A. (1996). Effects of cryopreservation on the viscoelastic properties of human arteries. Ann Vasc Surg..10 (3): 262-72. PMID:8792995

Sanders E.J, Prasad S. & Cheung E. (1988). Extracellular matrix synthesis is required for the movement of sclerotome and neural crest cells on collagen. Differentiation.Volume 39, Issue 1, November, Pages 34-41. PMID: 3246291

Schenke-Layland K, Madershahian N, Riemann I, Starcher B, Halbhuber KJ, König K, & Stock UA. (2006) Impact of cryopreservation on extracellular matrix structures of heart valve leaflets. Ann Thorac Surg. 81 (3): 918-926. PMID:16488695

Schenke-Layland K., Xie J., Heydarkhan-Hagvall S. Hamm-Alvarez, S. F., Stock U A., Brockbank K.G.M, & MacLellan W. R. (2007). Optimized preservation of extracellular matrix in cardiac tissues: implications for long-term graft durability. Ann Thorac Surg. 83 (5): 1641-1650. PMID:17462373

Schultz G.S., Ladwig G., & Wysocki A. (2005). Extracellular matrix: review of its roles in acute and chronic wounds. In:
   http://www.worldwidewounds.com/2005/august/Schultz/Extrace-Matric-Acute-Chronic-Wounds.html

Schwinn, M.K., Faralli, J.A., Filla, M.S., Peters D.M., (2010). The Fibrillar Extracellular Matrix of the Trabecular Meshwork. *Encyclopedia of the Eye*. Edited By Joseph Besharse, Reza Dana, & Darlene A. Dart. *Pages 135-141*. Elsevier.ISBN: 978-0-12-374198-1 USA.

Sottile J., Shi, F., Rublyevska, I.,Chiang, HY., Lust, J & Chandler, J. (2007). Fibronectin-dependent collagen I deposition modulates the cell response to fibronectin. *Am J Physiol Cell Physiol*. 293: C1934 – C1943. PMID: 17928541

Standards for Tissue Banking: International Atomic EnergyAgency (IAEA - 2005); *www.iaea.org*.Spanish Association of Tissue Banks (AEBT - 2005). www.aebt.org

Sweeney S., Orgel JP., Fertala A., McAuliffe JD., Turner KR., Di Lullo GA., Chen S., Antipova O., Perumal S., Ala-Kokko L., Forlino A., Cabral WA., Barnes AM., Marini JC., & San Antonio JD. (2008). Candidate Cell and Matrix Interaction Domains on the Collagen Fibril, the Predominant Protein of Vertebrates J. Biol. Chem. 2008. 283: 21187-21197. ISSN 0021-9258

Wolfe, J. & Bryant, G. (1999). Freezing, drying and/or vitrification of membrane-solute-water systems.. Cryobiology 39: 103 – 129. PMID: 10529304

Woo S.L., Orlando C.A., Camp J.F., & Akeson W.H. (1986). Effects of postmortem storage by freezing on ligament tensile behavior. J Biomech. 19:399–404. PMID:3733765

Worthley, DL. Giraud, A.S. & Wang, T.C.(2010) The Extracellular Matrix in Digestive Cancer

Cancer microenvironment official journal of the International Cancer Microenvironment Society. 3 (1) p. 177-185. PMCID: PMC2990481

Permissions

The contributors of this book come from diverse backgrounds, making this book a truly international effort. This book will bring forth new frontiers with its revolutionizing research information and detailed analysis of the nascent developments around the world.

We would like to thank Igor I. Katkov, Ph.D., for lending his expertise to make the book truly unique. He has played a crucial role in the development of this book. Without his invaluable contribution this book wouldn't have been possible. He has made vital efforts to compile up to date information on the varied aspects of this subject to make this book a valuable addition to the collection of many professionals and students.

This book was conceptualized with the vision of imparting up-to-date information and advanced data in this field. To ensure the same, a matchless editorial board was set up. Every individual on the board went through rigorous rounds of assessment to prove their worth. After which they invested a large part of their time researching and compiling the most relevant data for our readers. Conferences and sessions were held from time to time between the editorial board and the contributing authors to present the data in the most comprehensible form. The editorial team has worked tirelessly to provide valuable and valid information to help people across the globe.

Every chapter published in this book has been scrutinized by our experts. Their significance has been extensively debated. The topics covered herein carry significant findings which will fuel the growth of the discipline. They may even be implemented as practical applications or may be referred to as a beginning point for another development. Chapters in this book were first published by InTech; hereby published with permission under the Creative Commons Attribution License or equivalent.

The editorial board has been involved in producing this book since its inception. They have spent rigorous hours researching and exploring the diverse topics which have resulted in the successful publishing of this book. They have passed on their knowledge of decades through this book. To expedite this challenging task, the publisher supported the team at every step. A small team of assistant editors was also appointed to further simplify the editing procedure and attain best results for the readers.

Our editorial team has been hand-picked from every corner of the world. Their multi-ethnicity adds dynamic inputs to the discussions which result in innovative outcomes. These outcomes are then further discussed with the researchers and contributors who give their valuable feedback and opinion regarding the same. The feedback is then collaborated with the researches and they are edited in a comprehensive manner to aid the understanding of the subject.

Apart from the editorial board, the designing team has also invested a significant amount of their time in understanding the subject and creating the most relevant covers. They scrutinized every image to scout for the most suitable representation of the subject and create an appropriate cover for the book.

The publishing team has been involved in this book since its early stages. They were actively engaged in every process, be it collecting the data, connecting with the contributors or procuring relevant information. The team has been an ardent support to the editorial, designing and production team. Their endless efforts to recruit the best for this project, has resulted in the accomplishment of this book. They are a veteran in the field of academics and their pool of knowledge is as vast as their experience in printing. Their expertise and guidance has proved useful at every step. Their uncompromising quality standards have made this book an exceptional effort. Their encouragement from time to time has been an inspiration for everyone.

The publisher and the editorial board hope that this book will prove to be a valuable piece of knowledge for researchers, students, practitioners and scholars across the globe.

List of Contributors

Poh Chiang Chew and Abd Rashid Zulkafli
Freshwater Fisheries Research Division, FRI Glami Lemi, Jelebu, Negeri Sembilan, Malaysia

Qing Hua Liu, Zhi Zhong Xiao, Shi Hong Xu, Dao Yuan Ma, Yong Shuang Xiao and Jun Li
Center of Biotechnology R&D, Institute of Oceanology, Chinese Academy of Sciences, Qingdao, PR China

Zoltán Bokor, Béla Urbányi, László Horváth, Tamás Müller and Ákos Horváth
Department of Aquaculture, Szent István University, Hungary

Ofelia Galman Omitogun, Olanrewaju Ilori, Olawale Olaniyan, Praise Amupitan
Biotechnology Laboratory, Department of Animal Sciences, Obafemi Awolowo University (OAU), Ile-Ife, Nigeria

Tijesunimi Oresanya, Sunday Aladele and Wasiu Odofin
National Centre for Genetic Resources and Biotechnology (NACGRAB), Moor Plantation, Ibadan, Nigeria

Yusuf Bozkurt
Mustafa Kemal University, Faculty of Fisheries, Department of Aquaculture, Hatay, Turkey

Ilker Yavas and Fikret Karaca
Mustafa Kemal University, Faculty of Veterinary Medicine, Department of Reproduction and Artificial Insemination, Hatay, Turkey

R.K. Radha, William S. Decruse and P.N. Krishnan
Plant Biotechnology and Bioinformatics Division Tropical Botanic Garden and Research Institute Palode, Thiruvananthapuram, Kerala, India

Conceição Santos
CESAM & Department of Biology, University of Aveiro, Aveiro, Portugal

Marcos Edel Martinez-Montero
University of Ciego de Avila/Bioplantas Centre, Cuba

Maria Teresa Gonzalez Arnao
Universidad Veracruzana, Mexico

Florent Engelmann
IRD, UMR DIAPC, France
Bioversity International, Italy

Jiří Zámečník, Miloš Faltus, Alois Bilavčík and Renata Kotková
Crop Research Institute, The Czech Republic

David J. Burritt
The Department of Botany, The University of Otago, Dunedin, New Zealand

H. Perez Campos
Instituto Nacional de Donación y Trasplante (INDT), Ministerio de Salud Pública - Fac. de Medicina, Uruguay